Lecture Notes in Economics and Mathematical Systems 670

For further volumes:
http://www.springer.com/series/300

Jean-Pierre Aubin

Time and Money

How Long and How Much Money is Needed to Regulate a Viable Economy

Springer

Jean-Pierre Aubin
VIMADES (Viabilité, Marchés, Automatique et Décision)
Paris, France

ISSN 0075-8442
ISBN 978-3-319-00004-6 ISBN 978-3-319-00005-3 (eBook)
DOI 10.1007/978-3-319-00005-3
Springer Cham Heidelberg New York Dordrecht London

Library of Congress Control Number: 2013944195

Printed on acid-free paper

Springer is part of Springer Science+Business Media (www.springer.com)

This book is dedicated to Jean Cartelier, who, for a quarter of century, has guided my views on economics, by insisting on the integration of means of payments in a dynamical framework, and tried very hard to keep me away from "uconomia."

Foreword

Kreislauf des Geldes (circle of money) in Aachen
sculptured by *Karl-Henning Seemann*, 1977.

Time and *money*, rather, *duration* and *endowment*, are the themes scrutinized in this short book. Time and money are so polysemous concepts that we need to specify the meanings we shall use for diminishing the confusion that the different senses trigger: "duration" of temporal windows, fleeing with chronological time, during which is computed the endowment in units of numéraire. This endowment is needed for *regulating* the evolution of *economically viable commodities and prices* or financing transactions of commodity allocations during a *propitious temporal window*. Examples of temporal windows range from the centuries required for building cathedrals to the milliseconds of the high frequency markets. The computation of temporal windows motivates the introduction of *"duration functions"* with variable velocities chaperoning the evolution of economic variables (see Chap. 7, p. 105) for determining the propitious temporal windows on which they evolve.

"Regulation" means that we compute also the *transactions* of commodities (velocities of commodity evolutions) and the *price fluctuations* (velocities of prices) for allowing economic agents to satisfy dynamically (at each instant) their needs while obeying the scarcity constraints.

Without prices, economic viability imposes conditions on the transactions to barter commodities, to the risk of clashes and frustrations whenever the viable transactions are not *desired* by the economic agents, imposed to them or stolen or destroyed.

With prices (linear functions associating units of numéraire with commodities), economic viability requires an assumption on the joint evolution of transactions, fluctuations of prices, and units of numéraires: the sum of the *"transaction values"* and the *"impact of price fluctuations"* should be negative[1] or equal to zero. This assumption translates the *budgetary rule*[2] stating that *an agent cannot spend more in units of numéraire than what he earns after inflation*, still in units of numéraire. In an evolutionary world, this assumption implies constraints on the *transactions* and on the *price fluctuations* needed for regulating economically viable evolutions of commodities and prices.

When speculative transactions and price fluctuations determined by the crowd of economic agents, both consumers and producers, evolve inadequately, then either the economic viability is violated, or the budgetary rule is transgressed, or both.

The implicit current orthodox belief requires that budgetary constraint on (virtual) units of numéraire should be respected at the cost of violating (real) economic viability: hence the overall austerity, yet unfairly distributed, but that is another issue not dealt with in this study, since it is vigorously presented by *Joseph Stiglitz* in [120, Stiglitz] and his other books, where an abundant literature can be found.

The explicit proposition we raise is that the economic viability must be maintained, and, whenever it is at stakes, the budgetary rule should be violated in the following sense: *a strictly positive endowment in monetary units should be injected by an "endowment provider" (regulatory agency, central bank, "lender of last resort," etc.) in the economy for providing means of payment to the economic agents until economic viability is restored.*

If the endowment is negative, it is a profit for the provider, if it is null, the financial system is balanced, and if it is positive, the endowment is the number of units of numéraire needed to keep the economy going, correcting the prices for allocating commodities. Commodities are real, numbers are abstractions blending with reality. The endowment provider is the ultimate producer transforming units of numéraire into money, regarded as means of payments.

[1]They are also called, respectively, *commodities effect* and *price effect*.

[2]In a mythical static world, it is known under the *Walras law*.

Computing the minimum endowment restoring economic viability and deriving the dynamic laws regulating both transactions and price fluctuations is the problem that we shall attempt to solve by a mathematical metaphor.

In other words, we recover the missing hand of *Adam Smith*'s "invisible man," who guides the evolution of prices with one hand and the evolution of economic agents with the forgotten one. Hence, we suggest "him" to use a viable regulation map during an adequate (computed) "temporal window." The determination of the temporal window accompanies the specific viable endowment. This regulation excludes the fact that this endowment is created *ex nihilo*, since it constrains both the transaction of economic agents and the fluctuation of prices to keep the economy running.

Paris, France
November 2012

Jean-Pierre Aubin

Acknowledgments

I warmly thank *Alain Bensoussan* (University of Texas at Dallas), *Olivier Dordan* (Université Victor Segalen, Bordeaux), *Marie-Hélène Durand* (IRD, Institut de recherche sur le développement), *Efim Galperin* (UQAM, Université du Québec à Montréal), *Vladimir Lozève* (Natixis, Paris), *Frédéric Planchet* (ISFA, Institut de Sciences Financière et d'Assurances, Université de Lyon 1), *Jean-Philippe Terreaux* (IRSTEA, Institut de Recherche en Sciences et Technologie pour l'Environnement et l'Agriculture), *Richard Vinter* (Imperial College, London), *Jacques Weber* (CIRAD, Centre de coopération internationale en recherche agronomique pour le développement), and my "viabilist" colleagues and friends, among them *Luxi Chen*, *Anya Désilles*, *Nadia Lericolais*, *Sophie Martin*, *Patrick Saint-Pierre*, as well as *Hélène Clément-Pitiot*, for the discussions and the information gathered, and *Georges Haddad* for his permanent interactions on various topics. The numerical examples are due to *Patrick Saint-Pierre*.

This work was partially supported by the Commission of the European Communities under the 7th Framework Programme Marie Curie Initial Training Network (FP7-PEOPLE-2010-ITN), project SADCO, contract number 264735.

Organization of the Book

The organization of this short opuscule is bottom up, relegating technical mathematical details at the end. Any compromise between brevity and information is likely to be bad.

The viability concepts and theorems are presented in [15, *Viability Theory. New Directions*] and are not *duplicated* as much as possible: this short monograph offers some economic considerations which were not presented in this monograph by lack of space (except Sect. 15.3, p. 620), pursuing the quests exposed in [9, *Dynamic Economic Theory: A Viability Approach*], of which it is a continuation.

Chapter 1, p. 1, *The Underlying Thesis*, presents and comments the objective of this book (without mathematics) to explain the political economic motivations. We followed an inductive approach, climbing the long mathematical ascent toward abstraction. The hurried reader mainly interested by the final mathematical results may start with Chap. 6 and end with Chap. 7.

The next three chapters, devoted to "illustrations", state and present the concepts of endowment in the simplest possible framework. They will be adapted for computing the endowment functions to cover "fundamental values" in the Turgot sense and, next, to the economic "exchange values."

Chapter 2, p. 31, *How Long and How Much Endowing One Commodity*, studies the case of one commodity, one economic agent, and one endowment provider. This allows us to provide simple analytical formulas supplying right away regulation maps telling the provider what is the endowment to associate with the commodities to be financed. First surprise, this regulation map may be *set-valued*, i.e., produces *shocks*, in the sense of shock waves in physics: this simplest illustration leads to the fact that this regulation map is the solution to the *Burgers partial differential equation*, the cornerstone of fluid mechanics. In this context, endowment are constants.

We assume in Chap. 3, p. 47, *Keeping the Endowment Above a Viability Threshold*, that the provider manages the evolution of its endowment *for keeping it above a given time-dependent endowment threshold during a temporal window*. We shall provide the analytical formulas giving the initial value of the endowment (equity, solvency capital requirement, etc.) that the endowment provider should secure for keeping at each instant the endowment above the threshold.

In Chap. 4, p. 57, *Uncertain Endowments and Economic Cycles*, the rate of return is no longer fixed but has to be determined by regulation maps for guaranteing that the endowment remains viable between (constant) lower and upper bounds. This viability requirement restricts the choice of these rates and determines the set of regulation maps associating them with the endowment. This decor is sufficient to provide *viability oscillators* which, in this framework, illustrate *cyclic evolutions* of endowments, without assuming anything on the dynamics, the constraints being sufficient to trigger "economic cycles."

These chapters, dealing with "money" (shyly called "endowments" for avoiding unnecessary debates by using polysemous concepts), motivate the distinction between chronological time and temporal windows over which economic states evolve. It is time to devote Chap. 5, p. 67, *Evolutions and Their Temporal Windows*, to nonstandard considerations on time and duration, prediction and anticipation, and extrapolation from the past. Section 5.2, p. 70, *Duration Functions*, presents the results concerning duration functions, their fluidities, and the concepts of urge and haste. They will be used to chaperon the evolution of commodities and prices and to determine the temporal windows.

Chapter 6, p. 85, *Endowing Fundamental Values: Willingness to Pay*, is the most important one, since it studies the endowment functions of *Turgot*'s *fundamental value*, revisited nowadays under the name of *Willingness to Pay*. At this stage, for financing non-merchant economies, or organizations, biodiversity, etc., the concept of price linked to exchange economies is not relevant since ... there is no exchange and no market mechanism. Still, one can compute intrinsic values without mimicking the "Market" with virtual prices.

Chapter 7, p. 105, *Endowing Exchange Values: Adam Smith's Invisible Man*, invisible but endowed with two hands, the first one being the Adam Smith's hand guiding the evolution of prices, and the other one, hidden, advising the economic agents, is devoted to the endowment of *Turgot*'s *exchange* (or market) *value* when commodities are allocated among agents, (market) prices are introduced as well as duration functions chaperoning their evolution for determining the temporal windows. It is both a consequence and a generalization of the preceding chapter, a consequence since the variable x is now replaced by the duration functions chaperoning agent's evolution and the prices. The difficulty is to define these economic ingredients and in the complication of the formulas. Hopefully, the results of the preceding chapter provide the same interpretations, but in an economic framework more complex to describe.

Section 7.1, p. 105, describes the *Mathematical Framework of Dynamical Economic Theory* familiar to (mathematical) economists, maybe not for the readers discovering this field.

The rest of the chapter is devoted to the simplest example of endowment to this economic case for regulating viable evolutions in an economic environment depending both on chronological time and on duration functions. Many other examples involving more and more elements (individual means of payment, labor, capital, etc.) can be integrated at the price of difficulties of exposition. The programming of the viability algorithms to provide numerical examples remains to be done.

Chapter 8, p. 119, *Why Viability Theory*, describes the basic concepts and propose a *minimal viability survival kit* stating some of the basic theorems used in the second part of this short book.

Chapter 9, p. 127, *What Is to Be Done*, lists other urgent problems which should motivate further mathematical developments, not an exhaustive list, naturally, but open problems which could start to be studied with some of the tools of viability theory and, above all, to forge some new ones, which may and should be radically different. Among them, the tools presented below could be used to compute the insurance cost for a country hedging the *minimal level of wealth* guaranteed to each citizen. This minimal threshold should depend on many parameters (time, age, place, economic state, heath, etc.), a new way for defining social "classes."

These tasks are delegated to younger generations of mathematicians and economists to really collaborate by questioning the pages of this (and other) books, since human brains never cease to understand.

Contents

1 The Underlying Thesis 1
1.1 Investment Duration and Monetary Endowment 9
1.2 The Pantometric Trap 10
1.3 The Curse of the Exponential 13
1.4 Is the Shareholder Value a Value? 18
1.5 The Ockham Razor 20
1.6 Uconomia 23

2 How Long and How Much Endowing One Commodity 31
2.1 Provider Financing One Commodity 31
2.2 Restoring Viability 34
2.3 Providing Endowment 37
2.4 Viability Property of the Regulation Map 40
2.5 Providing Commodity 41
2.6 Towards Viability Solutions 42
2.7 The Viabilist Strategy 44

3 Keeping the Endowment Above a Viability Threshold 47
3.1 Deterministic Endowment Function 49
3.2 Viability Solution 49
3.3 Correction of Unviable Potential Functions 54

4 Uncertain Endowments and Economic Cycles 57
4.1 Inertia Principle and Heavy Evolutions 58
4.2 Viability Oscillators and Economic Cycles 63

5 Evolutions and Their Temporal Windows 67
5.1 Evolutions and Their Temporal Windows 67
5.2 Duration Functions 70
5.3 Maupertuis' Pleasure 74
5.4 Giving Duration Time 75
5.5 Evolutionary Systems 78

5.6 Anticipating from the Past or Predicting the Future? 80
5.7 Extrapolation by Historical Differential Inclusions 81

6 Endowing Fundamental Values: Willingness to Pay 85
6.1 A Static Viewpoint 85
6.2 An Evolutionary Viewpoint 86
6.2.1 The Canonical Endowment of Fundamental (or WTP) Value 87
6.2.2 The Viability Solution to the Endowment Problem 89
6.2.3 Optimality Theorems on Endowment Functions 90
6.2.4 Microscopic and Macroscopic Properties of the Endowment Function 92
6.2.5 The Endowment Effect 95
6.3 General Valuation Functions 96
6.3.1 Adding Other Costs to the Endowment 96
6.3.2 The Viability Solution 99
6.3.3 Optimality Properties 103

7 Endowing Exchange Values: Adam Smith's Invisible Man 105
7.1 A Mathematical Framework of Dynamical Economic Theory 105
7.2 Endowment Functions 112
7.3 Optimality Theorems on Endowment Functions 114
7.4 Microscopic and Macroscopic Properties 115

8 Why Viability Theory? 119
8.1 Viability Kernels and Capture Basins 119
8.2 A Viability Survival Kit 121
8.2.1 Bilateral Fixed Point Characterization 122
8.2.2 Viability Characterization 122
8.2.3 Tangent and Normal Cones 123
8.2.4 The Regulation Map 124

9 What Is to Be Done? 127
9.1 Insuring Each Citizen 127
9.2 Kingly Duties 129
9.3 Is It Time to Conclude? 130

References 133

Index 139

Index of Names 143

Chapter 1
The Underlying Thesis

Léon Walras cleverly designed an evolutionary system guiding "Adam Smith's invisible hand": the *Walras tâtonnement* is regarded as a mathematical translation of the law of supply and demand. It describes an a priori dynamical process governing the evolution of prices. If this evolution converges to a price, which is a (mathematical) *equilibrium*, it happens to be viable by design. However, *transactions are forbidden before reaching equilibrium*, so that the economic part of the system is never viable ("the markets are not cleared") before infinity, except when the initial price is a viable equilibrium. *"The long run is a misleading guide to current affairs. In the long run we are all dead. Economists set themselves too easy, too useless a task if in tempestuous seasons they can only tell us that when the storm is past the ocean is flat again"* wrote *Keynes* in the third chapter of hiss *A Tract on Monetary Reform* (1923).

Roughly speaking, *long run* investments are economic ones dedicated to the building of companies for designing, producing, and distributing real commodities and services to the agents, for rewarding the employees with real wages, etc. *Short run* investments are rather characteristic features of financial ones, where investors and speculators assess the financial characteristics of a company today *to form expectations on the speculative financial value in the future* during *auspicious* temporal windows on the basis of *volatile* company's share prices. Letting aside ethical, human, social, economical, and technical considerations, more and more (clever) perverse hijacks of financial tools[1] have been invented to evade the regulating role of central banks for clearing and settling accounts.

[1]For instance, dark pools of shadow finance, multilateral trading facilities, securization, collateralized debt obligation, credit default swaps, leverage by-outs, naked short selling, high frequency markets, etc. How this wonderful word "equity" came to mean the interest of shareholders (shareholders' equity) in a company and other stock owners? When returns on equity (ROE) measuring the rate of return on the ownership interest around 15 % and more became standards after the years 1980, were equities really equitable? Not being a native speaker English, I am quite reluctant to use in this context the concept of equity, which should only mean, as in other Romance languages, impartiality, fairness, justice, rightfulness, not to mention the concept of ethics. See

J.-P. Aubin, *Time and Money*, Lecture Notes in Economics and Mathematical Systems 670, DOI 10.1007/978-3-319-00005-3_1,

Economic investment profits all agents. One could avoid using finance, in principle, at least,[2] since barter transactions are feasible, but complicated to operate. However, as soon as numbers were invented, they were used to facilitate and lubricate[3] or catalyze[4] transactions, and, on the way, to profit mainly speculators. The emergence of numbers in social intercourse, including language, is a characteristic feature of what is called economics and finance to distinguish them from all other types of social relations.

Balancing the budget or cutting expenses may not solve the problem, whereas "endowment[5] providers" should create the means of payments in units of *"numéraire[6]"*. The word "money" is too polysemous to be used at this level. The definition of money given by Turgot, for instance, is the definition of "numéraire."

> **1** *[**Turgot's Money**] Every commodity has the two essential properties of money, of measuring and representing all values; and, in this sense, all merchandise is money.*

Among the roots of the problems lie at least two roles played by these units of numéraire, *wealth*, a commodity storing value for both instantaneous consumption and insurance of an uncertain future, and *means of payments*, a tool for regulating the economy. *Allocations of units of numéraire which can be exchanged between economic agents are regarded as commodities and subjected to scarcity constraints, whereas means of payment are provided to the economic system for allowing the transactions of commodities and services.*

---

among many other references [73, *Vingt propositions pour réformer le capitalisme*] by *Gaël Giraud* and *Cécile Renouard*.

[2]Proposition 12.2.11, p. 499, of [15, *Viability Theory. New Directions*] states that, using viability multiplier theory, closest "connection matrices" of links between agents to the identity matrix *regulate the same viable evolutions than the ones governed by prices* with minimal norm.

[3]For motor engines, oil is provided to lubricate the functioning of the engine powered by fuel. The endowment of oil should remain constant whereas fuel is consumed to power the engine. However, even for simple mechanical engines, oil endowment has to be modified during short crisis periods for avoiding the engine breakdown.

[4]For metabolic reactions, specific enzymes greatly accelerate their rate and are not consumed by the reactions they catalyze. However, enzymes may be inhibited or deactivated during the catalytic process.

[5]The word "endowment" is also polysemous. We shall mention the "endowment effect[35–96]" assumption. An *endowment fund* is a fund in which regular withdrawals from the invested capital finance specific purposes, too rarely proposed except by nonprofit organizations, universities, hospitals, etc.

[6]Numéraire, a term coined from ... metallic coins with a legal course since around 1720 and used during and after the French Revolution, is a basic standard by which values are measured.

Actually, we shall use the term "endowment provider" for it does not really mean what is understood as a "lender of last resort." The question is not to *lend* but to *provide* means of payments. Not a *credit* to be refunded in the future, even without interests, but *hic et nunc*, here and now, the *endowment* needed to lubricate the design, the production, the distribution, and the transactions of real commodities and services whenever prices and incomes are no longer financially viable. Not only to avoid the bankruptcy of banks or other financial institutions growing to be deemed "too big to fail," a reward for incompetence or dishonesty, but to provide economic and financial organizations, whatever their size, the means to succeed even *whenever the prices are no longer viable during a critical temporal window.*

The underlying thesis subjected to examination states that *the viability of the economic system induces financial constraints that should be accepted and not the other way around: complying with implicit ideologically enacted budgetary rules on these abstractions, they may and too often do jeopardize the viability of the economy and ruin it.*

Much has been said and written on this topic by so many talented economists, but little has been heard or read: among many other citations, [1, *Les cinq capitalismes: Diversité des systèmes économiques et sociaux dans la mondialisation*] by *Bruno Amable*, [70, *Dépasser le capitalisme*] by *Gérard Fuchs*, [73, *Vingt propositions pour réformer le capitalisme*] by *Gaël Giraud* and *Cécile Renouard*, [97, *End This Depression Now!*] by *Paul Krugman*, [99, *Pour éviter le Krach ultime*], [102, *Le livre noir du libéralisme*] by *Pierre Larrouturou* (preface of *Michel Rocard*), [100, *Capitalisme, désir et servitude*] and [101, *L'intérêt souverain*] by *Frédéric Lordon*, [113, *La Banque. Comment Goldman Sachs dirige le monde*] by *Marc Roche*, [115, *La démondialisation*] by *Jacques Sapir*, etc.

Our purpose is *to support this conjecture* by an original *mathematical metaphor*, justifying the definition of an endowment function of time, duration, commodities and prices for insuring the economic viability. These evolutionary concepts differ from those of static general equilibrium theory, which are based on other mathematical metaphors. *Different mathematics for analyzing different problems*. We devote a brief Sect. 1.6, p. 23 on "uconomia" to the legitimacy of mathematics in economics, to stress the fact that we are very conscious of the problem and very cautious in delimitating the validity of the statements.

The motivations are economics, the end results are mathematics.

For simplifying the presentation and the mathematical techniques involved, we do not take into account the private creation of money by credits, exchanging present time units of numéraire of a lender with a flow of future payments to his borrowers. They involve another facet of the concept of *duration* of the refunding schedule. This involves other mathematical treatments since, at each present time, agents have to subtract from their income the cumulated consequences of its past investments, requiring the use of historical evolutionary systems (see Sect. 5.7, p. 81). These technical treatments would deserve another whole book.

To be mathematically specific at this early stage of the exposition, we summarize the main problem addressed[7]:

2 ***[Mathematical Description of the Endowment Problem]*** *At each time t, we require that the pairs $(x(t), p(t))$ made of allocation $x(t)$ of commodities and of each price $p(t)$ are viable in the sense that they remain in a given economic environment $K(t)$. The sum $E(x(t), p(t))$ of*

1. *the* transaction value $p(t) \cdot x'(t)$ *of the transaction* $x'(t) := \frac{d}{dt}x(t)$, *velocity of the evolution* $t \mapsto x(t)$ *of commodities* $x(t)$;
2. *the* price fluctuation impact $p'(t) \cdot x(t)$ *where* $p'(t) := \frac{d}{dt}p(t)$ *is the velocity of the evolution* $t \mapsto p(t)$ *of prices* $p(t)$;

is called the impetus*: it measures the velocity*

$$E(x(t), p(t)) := p(t) \cdot x'(t) + p'(t) \cdot x(t) = \frac{d}{dt}U(x(t), p(t))$$

of the patrimonial value $U(x(t), p(t)) := p(t) \cdot x(t)$ *along the economic evolution* $t \mapsto (x(t), p(t))$ *of commodity-price pairs.*
The problem is to find the optimal temporal window $[T-\Omega, T]$ *over which we minimize the* maximal impetus $\sup_{t\in[T-\Omega,T]} E(x(t), p(t))$ *among all* economically viable *evolutions* $(x(t), p(t))$ *of commodity-price pairs, which is the* endowment $W(t, x(t), p(t))$ *requested to finance (or hedge, cover) the maximal impetus.*
The main problem studied in this book is to compute and extract the properties of the optimal temporal window and the optimal endowment $W(t, x(t), p(t))$.
The partial derivatives of the endowment function provide the "transaction rules" *to the agents for realizing their transactions and the price* "tâtonnement rules" *to the "Market" for taming the price fluctuations.*
If this endowment $W(t, x(t), p(t))$ *is not positive, then the financial or monetary dream is fulfilled.* Requiring it to be zero when it must be positive implies the collapse of the viability of the economic evolution. *Hence*

1. *the endowment provider knows the endowment function;*
2. *the agents their affordable transactions;*
3. *and the "Market" the economically viable prices.*

This information is more precise than the one derived from general equilibrium theory, because time and uncertainty are involved, *hence* transactions and price fluctuations, *not only commodities and prices as in the static case.*

[7]The minimization problem popping up below does not belong to the class of intertemporal optimization problems studied in optimal control theory: the integral criterion is replaced by a supremum criterion, which is not differentiable. The tools of viability theory allows us to overcome this difficulty.

This dynamic approach allows us to control both the transaction values of economic agents and the price fluctuation impact $p'(t) \cdot x(t)$ on a commodity $x(t)$. This is related to the concept of *inflation*. Inflation is the price fluctuation impact on a reference commodity, either a *numéraire* (the Gold Standard since King *Croesus* coined them out of Pactolus river in Lydia in around 610 BC and in China 600 BC until it was abandoned in August 15, 1971 under the presidency of *Richard Nixon*) or a *market basket* defined to be a *consumer price indexes* (CPI). Among them, is the nonofficial *Big Mac Index* (http://www.economist.com/search/apachesolr_search/big20mac20index), waiting for the apparition of the *Chow Mein Index*, 炒 面 指 数, although the *Fourastié Hair-Cut Index*[8] would provide a more judicious choice, since every man regularly confides his hair to a hair-dresser whose *hourly productivity is practically constant* in time and space. Only a tiny culturally biased minority of them are fed (up?) with Big Macs for their gastronomic and heath misfortune.

Agents are supposed to know how to realize their transactions. Living organisms, among which human *decision makers*, use whatever their cognitive systems for selecting their actions. They are here for three billion years and a half, long before us, maybe long after human beings, according to *Claude Lévi-Strauss*. They do not use numbers, they do not use neither probabilities nor statistics, they do not forecast, but they *anticipate*, another theme of this book.

Institutions are abstractions made of human brains and flesh which took over the possibility of *transforming present real economic and scarcity constraints into future abstract budgetary constraints*, thus, uncertain and virtual. Some are more greedy and cupid than others, more blind to the future than optimists inclined to believe positive forecasts. "After me, the deluge" (*"Après moi, le déluge"*), as Madame de Pompadour said when she perceived the warning signals of the approaching Revolution. "After me, the crisis," the lives of the citizens are the non redeemable loans of last resort, as many financial agents think.

Who thus can use the possibility to regulate the prices if it is not the subset of those economic agents who are also "price markets"? The "Market" (not to be confounded with the "markets," a code name under which speculators hide)?

Agents choose the prices of what they sell (including their labor) as they choose commodities, with the same ignorance of the underlying economic mechanisms, but with different punishments for this ignorance, some being undue rewards.

A common criticism of the role of lenders of last resort is that the backing of financial institutions is a temptation for them to take on more risk. Taking risks is the role of an entrepreneur who seeks to invent, produce, and distribute

---

[8] *Jean Fourastié* notable for having coined the expression Trente Glorieuses ("the glorious thirty [years]") to describe the period of prosperity that France experienced from the end of World War II until the 1973 oil crisis (1945–1973), proposed in 1950 to use men's hair-cuts as such an index: *"A specimen Paris hairdresser did not'produce' more hair-cuts in 1950 than in 1900; the number was 3.5 hair-cuts per working hour in both periods. [...] the 'consumption' of hair-cuts has increased more than that of wheat, although the productivity of hairdressers has remained stationary, while that of wheat growers has sharply increased."*

commodities and services under uncertainty. This is different from the risks taken by financial institutions[9] for producing monetary units by trading property *rights* and/or creating money from money for money, producing numbers made of loan interests or speculation on financial assets. Dangerous abstract numbers, since they can be transformed in real wealth without real economic counterpart. Even at the theoretical stage, there is no explanation on how the "Market" can avoid this confusion of roles.

Instead, the *ideological criticism* of a lender of last resort is that its existence reveals the possibility of the failure of the "Market": if "He" accurately assesses prices and risks, then institutions will not misuse the capital and the idea of a panic or "mimetic" credit crunch spreading through the banking system would be impossible.[10]

The faith into the power of this new (undefined) deity, the *"Market"*, knowing best how to "regulate the economy" with an *"invisible iron hand"* for the multitude, a *golden* one for a few. "He" alone is capable of producing and allocating commodities and services to guarantee a population to survive, or, in the best cases, live. Their worshippers share the conviction that their deity can anticipate, calculate, or dictate prices, especially without regulations by political interventions other than the ones which forbid to prohibit them. They travestied the wonderful word of "freedom" by qualifying "free economy" an economy providing free lunch for the "markets," an alias for plain speculators[11] and modern rentiers, and free from any government regulation by lending them *with interests* predatory credit to be refunded by the taxpayers, instead of asking the political authorities, on behalf of their citizens, to "provide" the adequate endowment, with no interest.

The "Market" is rather a Ponzinian[12] deity unable to curb the exponential growth of the cost of loans awarded to debtors, themselves invisible thanks to the "securization" mechanisms.

However, the fear of the *Greenback Mill* is still alive, although one should understand how it operates to cure financial disorders.

---

[9]The risk issue is not dealt with in this text, but is the theme of a companion book [17, *Tychastic Measure of Viability Risk. A Viabilist Portfolio Performance and Insurance*] devoted to the financial system, whereas this one is devoted to the economic one for avoiding confusion by mixing problems.

[10]See, among many references, [41–46, Cartelier].

[11]Who have succeeded where the alchemists failed in turning paper into gold, who invented perpetual debt where physicists had to abandon the perpetual movement, and operated at their exclusive benefit the "exuberance" of the exponential (see Sect. 1.3, p. 13).

[12]Named after *Charles Ponzi*, who rediscovered a variant of pyramid form of fraud relying as on a mistaken belief in a nonexistent financial reality, including the hope of an extremely high rate of return, that pays returns to its investors from their own money or the money paid by subsequent investors. In 2008, Bernard Madoff used a similar scheme that cost his investors about 53 times the losses of Ponzi's scheme.

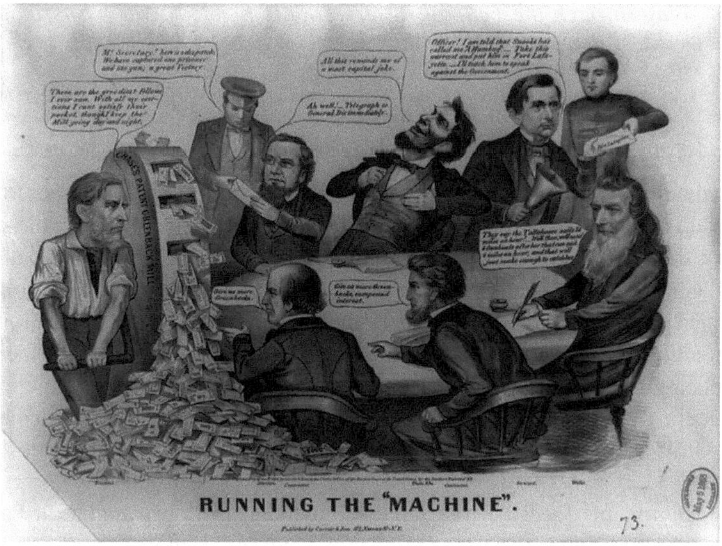

3 ***The "Greenback Mill".*** *A 1864 scathing attack against* Lincoln *who financed Civil War, created the Bureau of Engraving and Printing of the US Treasury Department, which issued paper money (Legal Tender Note) from 1862 to 1971: the "Greenback Mill" was turning out paper money by the million.*
The only available way to finance the war when he took office was to ask high-interest bonds to the Associated Banks of New York, run by British agents. Lincoln refused to pay these interests, and, on the advice of Colonel Edmund Dick Taylor, *the "father of the greenback", signed the First Legal Tender Act in 1862, followed by the National Banking Act of 1863 asking the Treasury to issue several hundreds of millions of new "greenbacks." This Greenback Mill financed not only the Civil War, but the construction of the first transcontinental railroad, grants of land and subsidies before he was assassinated on April 14, 1865.*
He followed the lessons of George Washington *who wrote in 1797 to his nephew that* "There is no practice more dangerous than that of borrowing money; for when money can be had in this way,' repayment is seldom thought of in time, the interest becomes a moth, exertions to raise it by dent of industry ceases, it comes easy and is spent freely, and many things [are] indulged in that would never be thought of if [they were] to be purchased by the sweat of the brow [. . .] in the mean time the debt is accumulating like a snow ball in rolling."
After Nelson Aldrich *and executives representing the Rothschild's, Lazard Frères, Israel Sieff, Kuhn Loeb Company, Warburg Company, Lehman Brothers, Goldman Sachs, Rockefeller and J.P. Morgan met for ten days at Jekyll Island (Georgia) in November 1910 in secret, from both the public and the government, the Glass-Willis proposal was hotly debated before President* Woodrow Wilson *signed on December 23, 1913 the Act creating the privately owned* Federal Reserve Bank.
The FED creates money, then loans it back to the government charging interest, the government levying income taxes to pay the interest on the debt. By 1928, the FED had taken over much of the currency market.
On June 4, 1963, John Kennedy *signed executive order 11110 to give the Treasury Department the Constitutional power to create and issue currency (money)*

without going through the Federal Reserve Bank. Kennedy was assassinated on November 22, 1963. Since 1968 they have been redeemable only in Federal Reserve Notes and are thus obsolete.

At the extreme left the Secretary of the Treasury complains that he cannot satisfy the greed of the contractors for "more greenbacks" (see [40, *Lincoln Money Martyred*] by *Butler*. Digital ID: http://hdl.loc.gov/loc.pnp/cph.3a11882)

The contemporary Golden Calf encourages speculators to sell "short" or to insure that they do not have (Credit Default Swap, for instance), to gamble with the savings of others since the fall of the barriers between investment and commercial banks (as the "Glass-Steagall Act" voted in 1933 and repealed in 1999). Wizards who encouraged powerful borrowers to borrow themselves, but with rates that allow them to repay simultaneously their debts and profit. Mage putting "levers" into their hand for producing perverse "effects" that increase their financial wealth at the expense of producers, for reducing also the cost of labor to increase productivity in their sole benefit, while requiring taxation laws more favorable to them than to the rest of the population.

Too many agents feed the monetary endowment, from private credit banks to central banks and States. The choice of constraining the monetary endowment is political: States and/or Markets. None of them is perfect. The kingly functions of States are designed to protect their citizens from wars, crimes, etc., and more and more, social, economic, and financial viability. Taxpayer money and sovereign debts are designed to finance the States for protecting the citizens. In democracies, the citizens can control in theory the decisions of civil servants, but not economic power of short-term cupid interests, which disappear only when they bankrupt. The problem is to understand how to select the crowd of decision makers at all level of the society and check their performance, at the lowest level, as the subsidiarity principle recommends. Hence the problem we face is *a problem of coordination of agents deciding their transactions and, for some of them, making prices.* There is not yet, at our knowledge, convincing mathematical treatment of this problem, even at level of schools of fishes or Panurge sheeps.

Are the flaws of the market economy caused by too much government regulation? The so-called inherent inefficiency of planning procedures due to the incompetency of planners (who exist since the dawn of time, and, nowadays, both in government institutions and in large private companies for devising Gosplans and business plans) as well as the criminal drifts of greedy financial agents are not sufficient explanations of the dysfunctions of the economic systems. Both are due, probably, to the frightening "mystery of the duality made of power and wealth." Ambitions use power to enrich themselves and/or enrich themselves to exercise power, generation after generation. Vested financial (compounded?) interests have embezzled the original (and moral) concept of "free market," from those of *The Theory of Moral Sentiments* by *Adam Smith* to the ones of *Recherche de l'idéal social*, title of a book by *Léon Walras* among many other ones.

The fears of *John Galbraith* of *private affluence and public squalor* that he mentioned in his 1888 [71, *The Affluent Society*] were optimistic: over the past 40 years, most people have seen only modest salary increases: the "income fault," as the San Andreas fault, widened so fast that we know that "the next big one" will happen, but not when. The crises of 1929 and 2008 are the consequences of weak States vis-à-vis the Markets. Are not Jefferson, Lincoln, Roosevelt, Kennedy who wants, at the very least.

The language is often used to replace missing objects or events by invented concepts, or divert them from their meanings, becoming new objects of the cultural milieu. Thus we never invoke morality as often as it is violated! Thus neo-liberalism of our day has gone against the core of the classical concept of free market to defend inequality of opportunity for certain economic agents and for their own exclusive benefit, as bureaucrats and apparatchiks have diverted state powers to monitor citizens of democracies as advanced as popular. Or our modern Eurocrats who evoke the subsidiarity principle, whereby decisions are be taken at the lowest level, to better camouflage their bureaucratic definitions of cheese, chocolate, and European jams!

1.1 Investment Duration and Monetary Endowment

Duration and fleeting time perceptions are different from person to person, as well as their desires and abilities. They lead to a "division of durations," as division of labor. The division or labor, which is at the root of the trade and economic interdependence, was recognized from *Plato* to *Adam Smith* through *Duhamel du Monceau* and *David Hume*.

Time and durations are involved in the trade of promises, which, for lack of trust in human pledges, is at the root of speculations. *Trust is not measurable*, and should escape from economics and mathematics and left to the sagacity of cognitive scientists. It plays a central role in human decision mechanisms, and thus, in economics.

Is the trend to shorter and shorter investment durations justified for adapting economic and financial mechanisms to production and distribution of commodity and services, or plain short-sighted greed? Is the craving of human brains for higher and higher velocities, including the "velocities of durations," describing the feeling of *"fluidity of time"*, justifies the inclination for "liquidity" of certain *commodities allowing instantaneous transactions*, such as various kinds of numéraire and money?

Time and money, together, arise as the main features that evolutionary economics and finance have to deal with.

Not that *time is money*, as *Benjamin Franklin* suggested in his *Advice to a Young*: *"Remember that TIME is Money."*, a short evolutionary comment. He added: *"In the Colonies we issue our own money. It is called Colonial Scrip.* We issue it in proper proportion to make the products pass easily from the producers to the consumers.

In this manner, creating ourselves our own paper money, we control its purchasing power, and we have no interest *to pay to no one.".*[13]

Nor that *money is time*, although it buys it, as an insurance against the anguish of an unforeseeable future, or a lack of confidence in the promises and claims of good faith.

But *time **and** money*, rather, *duration **and** monetary endowment*, evolving together, duration "chaperoning[14] money," so to speak. Because *capital* is, in the last analysis a, duration, a kind of *"frozen temporal window" of deprivation* (see [35, *History and Critique of Interest Theories*], by *Eugen Böhm von Bawerk*, a penny pinching libertarian).

Past, present, and future are involved. Past, for accumulating imperishable commodities without consuming them, present, for consuming part of this accumulation and saving or lending the rest for future purposes, including the reward of earlier deprivation, betting on faith and confidence. Past, present, and future do also evolve, the future turning into present to be instantaneously transformed into past.

Duration of deprivation, illusory, and gambling desire to forecast the future, trust in the reliability of the commerce of promises, instantaneous decisions, etc. play a crucial role in the functioning of the economic machine. For it requires some duration for producing and distributing commodities, using available technological and scientific knowledge. At this level, too, *increasing velocities of the products of cultural knowledge* play an important role.

1.2 The Pantometric Trap

As soon as exchanges became too complex to be operated by plain barter, various "numéraires" were created. The number of their units was the prototype of a concept of "value." Not only human brains give themselves up to forecast the future instead of anticipating it (see Sect. 5.6, p. 80), but they used numbers to "measure" ... anything, and not only "things" endowed with an explicit unit of measure to be

[13]The English bankers, being informed of that, had a law passed by the British Parliament prohibiting the Colonies from issuing their own money, and ordering them to use only the gold or silver debt-money that was provided in insufficient quantity by the English bankers. The circulating medium of exchange was thus reduced by half, triggering the Revolutionary War against England, followed by the Declaration of Independence in 1776 and the American Constitution, signed at Philadelphia, in 1787, which stipulates that "Congress shall have the power to coin money and to regulate the value thereof" in article 1, Sect. 8, paragraph 5. It was the beginning of the still existing recurring conflict between the US treasury and the privately owned Federal Reserve Bank created in 1913.

[14]As, in molecular biology, when *Ronald Laskey* called in 1978 "molecular chaperones" proteins assisting other macromolecular structures to fold and assembly, or as old aunts were chaperoning their young and temporarily innocent nieces in older times.

measured. The measurements are called "values," a polysemous word. Doing so, these values fall in the "pantometric" trap:

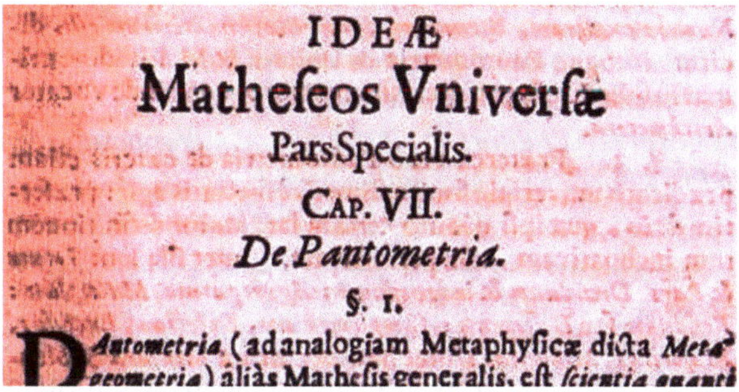
IDEÆ
Matheſeos Vniverſæ
Pars Specialis.
CAP. VII.
De Pantometria.
§. I.
PAntometria (ad analogiam Metaphyſicæ dicta *Meta-geometria*) aliàs Matheſis generalis, eſt *ſcientia quanti*

4 [**Quantitative Versus Qualitative Evaluation?**] *Friedrich Hayek* wrote in [86, *The Fatal Conceit*] that *"The belief derived from physics that measurement is an essential foundation of all sciences is very old. There was more than 300 years ago a German philosopher named Erhard Weigel who strove to construct a universal science which he proposed to call Pantometria, based as the name says on measuring everything.*
Much of economics, and if I may add in parenthesis much of contemporary psychology, has indeed become Pantometria in a sense in the principle that if you don't know what measurements mean, measure anyhow because that is what science does." (see [126, *Ideæ matheseos universæ cum speciminibus inventionum mathematicarum*]).

Cognitive sciences do not sustain that rationality of human brains can be reduced to *the maximization of utility functions*, the existence of which was already questioned by *Henri Poincaré* in a letter to *Léon Walras*: *"Satisfaction is thus a magnitude, but not a measurable magnitude."*

Blessed are the bonobos, which do not talk nor count. Alas, languages and numbers allowed men to throw them out of the little paradise that remained for them. If Aesop had known Pythagoras, he would have amended his famous trick offering dishes of tongues to Xanthus by adding digits, and explain *"The tongue and the digit are at once the best and the worst entertainment."* Numbers then became mankind's pantometric drug.

In trying too hard to serve, *value does not exist*, unless it appends a long list of adjectives qualifying what we mean when we deal with this concept. The task before us requires that we compile this list of adjectives. As soon as a meaning is chosen, while mingling with all, the concept of value defies analysis beyond any exhaustive classification.

5 ***[The Value Does Not Exist]*** *There is no concept of value as long as we do not specify precisely for instance,*

1. *which is evaluated (subsets and/or their elements);*
2. *by which to evaluate it (choice of the space of values);*
3. *by which procedures (valuators) a classification and/or assessment is operated;*
4. *for what and to whom this is evaluation useful;*
5. *etc.*

In this book, the value space is still the space $\mathbb{R}$ of real numbers. Another choice could be the space of subsets of a set, providing qualitative information rather than quantitative measure (see [60, Dordan] and the forthcoming [23, *Evaluation and Quotations of Sets*] by *Olivier Dordan* and the author).

Hence real numbers were and still are used for measuring commodities and services when they are endowed with units of measure. For that purpose, a numéraire has to be chosen, for prices to associate with commodities their value in units of numéraire. Nowadays, dollars, euros, yuans, yens, etc., waiting for a planetary numéraire, thrive to become the ultimate measures of almost all things (see for instance, [29, Money, Sex and Happiness: An Empirical Study] by *David Blanchflower* and *Andrew Oswald*).

Already, we face at least two concepts of economic value, fundamental and exchange values, as *Turgot* stated in his letter[15] to *Hume*:

6 ***[Letter from Turgot to Hume [1767]]*** *"Two types of value may be distinguished:* fundamental value *and* exchange value. *The fundamental value is what the thing costs to him who sells it, that is, the raw material cost, the interest of the advances, the wages of labor and industry. The exchange value is the price which the buyer agrees upon with the seller. The fundamental value is fairly stable*

[15]**Lettre de Turgot à Hume :** *La valeur fondamentale est ce que la chose coûte à celui qui la vend, c'est-à-dire les frais de la matière première, intérêt des avances, salaires du travail et de l'industrie. La valeur vénale est le prix dont l'acheteur convient avec le vendeur. La valeur fondamentale est assez fixe et change beaucoup moins que la valeur vénale. Celle-ci ne se règle que sur le rapport de l'offre à la demande ; elle varie avec le besoin, et souvent la seule opinion suffit pour y produire des secousses et des inégalités très considérables et très subites. Elle n'a pas une proportion nécessaire avec la valeur fondamentale, parce qu'elle dépend immédiatement d'un principe tout différent ; mais elle tend continuellement à s'en rapprocher, et ne peut guère s'en éloigner beaucoup d'une manière permanente. Il est évident qu'elle ne peut rester longtemps au-dessous ; car, dès qu'une denrée ne peut se vendre qu'à perte, on cesse de la faire produire jusqu'à ce que la rareté l'ait ramenée à un prix au-dessus de la valeur fondamentale. Ce prix ne peut non plus être longtemps fort au-dessus de la valeur fondamentale, car le gros prix, offrant de gros profits, appellerait la denrée et ferait naître une vive concurrence entre les vendeurs. Or l'effet naturel de cette concurrence serait de baisser les prix et de les rapprocher de la valeur fondamentale.*

and changes less frequently than the exchange value. The latter is ruled by supply and demand, it varies with needs, and often a single event suffices to produce very considerable and very sudden fluctuations. It is not in any essential proportion to the fundamental value, but it has a tendency to approach it continually, and can never move far away from it permanently. It is obvious that it cannot remain below it for a long time; for, as soon as a commodity can be sold only at a loss, its production is discontinued until the resulting scarcity has again raised it to a price above its fundamental value. The price can similarly not be much above the fundamental value for any length of time, for this high price, implying high profits, would call forth the commodity and generate lively competition among the sellers. Now the natural effect of this competition would be to lower the price until it again approaches the fundamental value." (See [75, The Turgot Collection*].)*

The least we can say is that there is no consensus on the role of money as a "value." But durations can be measured, since clepsydra and hourglasses were designed to do so, as an objective value (see Chap. 5, p. 67). *Adam Smith* stated that the *interest rate is the time cost of money*. *Karl Marx* recognizes in the first pages of *The Capital* the crucial role of "labor time" in the definition of "labor value":

7 ***[Karl Marx's Labour Time Value]*** *"The value of one commodity is to the value of any other, as the labour-time necessary for the production of the one is to that necessary for the production of the other. [...] In general the greater the productiveness of labour, the less is the labour-time required for the production of an article, the less is the amount of labour crystallized in that article, and the less is its value; and vice versa, the less the productiveness of labour, the greater is the labour-time required for the production of an article, and the greater is its value. The value of commodity, therefore, varies directly as the quantity, and inversely as the productiveness, of the labour incorporated in it."*

These two aspects of economic values, the *fundamental* and the *exchange* value, will be treated successively in this book.

1.3 The Curse of the Exponential

Then arose the issue of coevolution of the value of the numéraire available to economic agents, *financial value* created by the credit, "real" *economic value* created by labor for producing goods and services.

Compensating for privation of wealth during a given period is necessary, since we all use financial values to invest for *future* production and sell our work during the investment temporal windows we devote to it. The question is: *at what price?*

If *one accepts that the ultimate value is duration*, the duration of savings and the duration of work should keep pace, comparing deprivation and labor for one to offset the other.

Their values should evolve at the same rhythm, not only for reasons of equity, but for consistency and viability for the financial income to match the one derived from work, production and distribution of goods. This is not necessarily the case when the use of simple interest, calculated in proportion to the principal, has been replaced by the use of compounded interest, where the interest is added to the principal ("a double charge," according to Saint Thomas Aquinas, both the charge of a thing and the charge of its use, the "usury," from the Latin *usura*, profit derived from money lent, meaning nowadays an excessive interest rate).

If simple interests increase the financial value in proportion to the duration, compounded interests imply an exponential (or geometrical) growth of the financial value, which increases well beyond that of labor and production. To guarantee the viability of the coupling of the financial systems (without scarcity and immediate visible constraints other than the inertia thresholds of human beings), and economic systems (subject to vital constraints on human beings and scarcity of services and commodities), it is mandatory that financial and economic values evolve at the same rhythm to prevent the imbalances, the accumulation of which leads to catastrophes known as "crises."

Because the roles are not symmetric between lenders and borrowers. The poorer are the borrowers, since they cannot survive without borrowing *immediately*. They do not enjoy the choice that capital owners have to *guarantee duration and insure uncertainties of life*, who have an incentive to increase rates for enhancing their future while impoverishing borrowers. The extreme wealth of the first ones fuels the extreme poverty of others, in much greater numbers. This system is condemned as parasites that kill organisms they colonize, and which, consequently, disappear with them.

However, in addition to their short-sighted greed, the "creators of credit" refuse sternly to lower their interest rates and lessen their power by shrinking incomes of their loans. They corrupt other powers, both political and ideological, to prevent them to prohibit usury and deregulate finance into shadow finance, and to encourage the poor to run into debt. Especially when these potential victims are communities, companies, or states: by resorting to the "too big to fail" blackmail threatening that their own loss causes the losses of the communities, they take into their torment their employees and citizens, whereas the contrary happens, later, maybe, but worse. The surge of this infernal machine without brakes set to benefit lenders is a cause of business cycles.

The "curse of the exponential" is in the last analysis the ultimate key opening economic crisis, when the interests of the debt, increasing both autonomously and exponentially, drain on revenues of production and distribution of commodities and services to repay creditors, since the dishevelled growth of financial value is incompatible with that, slower, of economic activities operating on their own rhythm. Financial growth thus becoming incompatible with that of economic progress, triggers inflation, or an economic slowdown fueled by austerity policies which are counter-productive in that they worsen the situation, or *a restructuration of the excessive interest of a debt that has ceased to evolve in harmony with the value of goods and services.*

For instance, to underline the *unviable* exuberance of exponential evolutions obtained with fixed rates obtained by constant interest rates, we observed the daily rates of the CAC-40 index during a 2-year period, their average and the EONIA riskless rate. We summarize our observations in the following two figures.

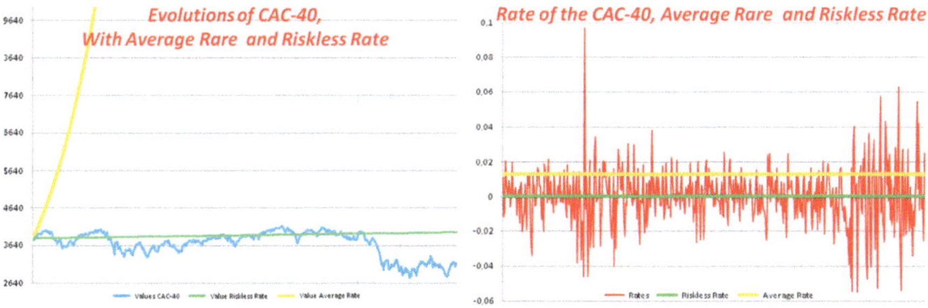

8 *[**Evolution of the CAC-40 Index and of its Returns**][**Left**]This figure displays from bottom to up the evolutions of*

1. *the CAC-40 index during 517 dates ending on December 09, 2011;*
2. *value governed by EONIA riskless rate from the index at initial time;*
3. *the value governed a constant equal to the average of CAC-40 rates from the index at initial time.*

*[**Right**] The figure displays the evolution of the daily rates of the CAC-40 index, the average daily CAC-40 rate during this period (0.01289) and the daily riskless rate (0.00008). The minimum of the daily rates of the CAC-40 index is equal to 0.000952 and the maximum to 0.02482.*

Actually, we know the upper bounds (*HIGH*) and lower bounds (*LOW*) of the CAC-40 index provided by brokerage firms. They delineate a tube in which lie the closing values of the index. The figure below displays this tube and the derived tube of returns taken during a period of 75 days in 2012. No exponentials pop up at first sight.

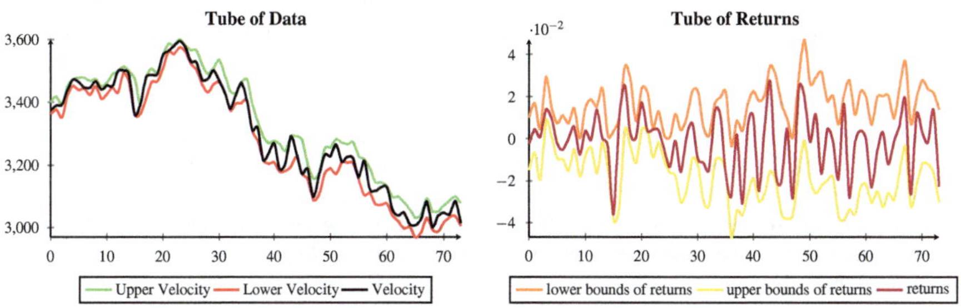

More rigourously, "viability detectors" provide evolutions governed by a dynamical system viable in a tube around a time series, provides the temporal windows during which the evolution remains in this tube. The fewer the temporal windows and the longer they are, the better is the detection process of recognition of the time series. In this case, it hints at *laws explaining* the evolution of the time series, providing more information than pattern recognition mechanisms which may reproduce patterns (such as statistical models, interpolation by spline functions, the VPPI extrapolator, etc.) without providing interpretations of the phenomenon involved, if any.

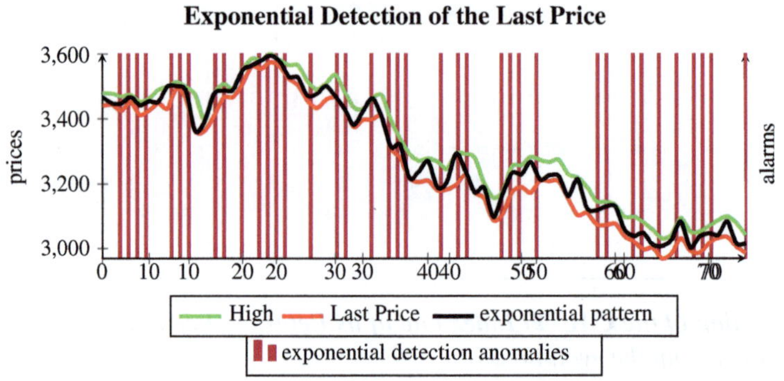

9 [*Exponential Detector*] *Here, the differential equation detecting patterns is a linear system providing exponentials patterns. The temporal windows are indicated by bars. The figure provides an example indicating that exponential financial growth provided by compounded interests does not fit the evolution of the CAC 40 index in its price tube since the detected windows are very small.*

Actually, one can prove that the stock market asset tubes are not viable under the geometric models, deterministic or stochastic, used almost universally in mathematical finance (as the simplest one leading to the Black and Sholes formula for estimating the prices of European options), which are not validated by experiments right at the beginning (see [17, *Tychastic Measure of Viability Risk. A Viabilist Portfolio Performance and Insurance Approach*]).

A cultural gap separated the physicists[16] (who, following Isaac Newton, are using the velocity and the acceleration to design "evolutionary engines"), and economists, favoring "growth rates" (which prohibit evolutions other than exponential growth

[16]And chemists, as the Nobel prize *Frederick Soddy* who wrote four books on money, including his [117, *The rôle of Money*] published in 1934, in which he underlined that financial debts grew exponentially at compound interest but the real economy was based on exhaustible stocks of fossil fuels, anticipated some reforms such as the floating of exchange rates, etc., and coined the word *ergosophy*, leading to the present day energy economics. I thank *Hélène Clément-Pitiot* for having mentioned this reference.

that feeds on itself). Velocities lead to simple interest rates, growth rates trigger compounded interests.

The expansion of goods and services, creators of "real value" should be the counterpart to the virtual "financial value." Both are expressed by numbers, but do not mean the same thing. Polysemous, the word loses sense: the *creation of shareholder value* (see Sect. 1.4, p. 18) goes in par with the still silent *destruction of production value* in the absence of regulation and "Lamarckian" compartmentalization devices to curb this growth for *guaranteeing the viability of a system by coordinating the viability of its components*.

It is thus impossible that *a mechanism based on compounded interests instead of simple ones can "stabilize" (a polysemous word) by itself, without regulation from institutional (governmental) authorities*[17] *prohibiting usury*.

Is there a mathematical justification of the fact that the velocity of economic evolution is proportional to its quantity, excluding, ironically, the impossibility of nonzero equilibrium, the Grail of economic *doxa*?

Known since the dawn of history[18] (see for instance [63, Einzig], [87, *An Ancient Economic History*] by Heichelheim and [89, *A history of interest rates*] by Homer and Sylla), compounded interests were regarded as the worst kind of usury and condemned by all moral institutions, from the Bible, under the name of the Hebrew *ribbit*[19] to the Gospel[20] and the Koran (Quran), under the name of *riba*.[21] It was also condemned by many philosophers, among whom Plato, Aristotle, Cato, Cicero, Seneca, etc., as well as by Roman law and the common laws of many other countries.

---

[17]Napoléon clearly understood the root of Europe's problem. For, after seeing a table of interest charges, he commented: *"The deadly facts herein revealed, lead me to wonder that this monster, interest, has not devored the whole human race. It would have done so long ago if bankruptcy and revolutions had not acted as counter poisons."* according to the books [65, *The Economic and Social Problems*] by *Michael Flürscheim* and [40, *Lincoln Money Martyred*] by *Butler*. However, the four Rothschild brothers used this deadly weapon to defeat him in Waterloo by building a network of agents shipping millions of coins to the Duke of Wellington, together with an "information mix" of true news and false rumors all over Europe, as a private intelligence service (subsidizing also the German, Austrian, Belgian, and Russian allies) who warned *Nathan Rothschild* the news of Wellington's victory at the Battle of Waterloo by optical telegraph a full day ahead. Hence by mimetic behavior other panicked investors sold their savings at their lowest price before the official news. Then *Rothschild* bought what he had just sold, and is fortune was multiplied twenty times on that day when misfortune sent Napoléon to Saint Helen to meditate on what would happen had he won in a world without compounded interests nor speculation!

[18]Under the Code of Hammurabi, one could hypothecate his wife. But a creditor could seize her for only 3 years and would have to return her in as good condition as she came.

[19]"Unto a foreigner thou mayest lend upon interest; but unto thy brother thou shalt not lend upon interest; that the Lord thy God may bless thee in all that thou puttest thy hand unto, in the land whither thou goest in to possess it." (Deuteronomy 23:20).

[20]"Why then didn't you put my money on deposit, so that when I came back, I could have collected it with interest?" (Luke 19:23)

[21]"O ye who believe! Devour not usury, doubling and quadrupling (the sum lent). Observe your duty to Allah, that ye may be successful" (Quran 3:130).

1.4 Is the Shareholder Value a Value?

The "shareholder value model" states that the *ultimate measure of a company's success is the extent to which it enriches shareholders*, and thus should be maximized. It became a growing compulsion after August 12, 1981, when *Jack Welch*[22] gave a talk intituled *"Growing fast in a slow-growth economy"*, a landmark on the silent revolution giving a lethal stop to the American *New Deal* for returning to the years of wealth and excessive speculation preceding the Wall Street Crash of 1929.

Silent, because no Bastille nor Winter Palace was assaulted, no preconceived secret coups were plotted. It was slow, and thus imperceptible, and democratically hatched by the majority of those who were in their twenties 20 years earlier in several countries, in the lapse of few years, from California to Poland through Paris in May 1968, conducting what appeared in retrospect a shift to individualism. This generation and the next ones abandoned the welfare society and the management processes that have been invented to cure the 1930 crisis for half a century. It traded a sense of qualitatively assessed collective solidarity for a frenetic quest of individualist success quantitatively measured by money or graded by rates. Ironically, it replaced individual responsible decision-making processes by irresponsible collective ones hidden in opaque committees.

The shareholder value took preeminence over the welfare of the workers and the production and distribution of commodities. Finance speculation divorced from economic realization, and massive deregulation advocated at that time triggered the opening of unbridled and unjustified income gap. *A new perception of time increased its fluidity* while narrowing the duration of temporal windows, the pressures of short-term financial gain at the expense of industrial investment. Short-term and instinctive agitation to the detriment of long-term actions which are triggered after adequate time maturation, were accelerated by the contemporaneous Digital and Internet Revolution. However, the human brains who created those faster and faster computers with larger and larger memory are the same than the ones of our ancestors 200,000 years ago.

However, in March 12, 2009, *Jack Welch* said that *"On the face of it, shareholder value is the dumbest idea in the world. Shareholder value is a result, not a strategy [...] your main constituencies are your employees, your customers and your products.*[23] The same article quotes *Alan Greenspan*, the former chairman of the Federal Reserve and a "high guru" of *laisser-faire* capitalism, told that *"the US might have to nationalize some banks on a temporary basis to fix the financial system"*. Although much too late, their (courageous and honest) confessions replaced the lines already written by the author before their recent unbosoming came to his knowledge.

---

[22] Welch became in 1981 the youngest chairman and CEO General Electric and has been nominated in 1999 the "manager of the century"!

[23] See http://www.ft.com/intl/cms/s/0/294ff1f2-0f27-11de-ba10-0000779fd2ac.html#axzz1eiLpL2PZ.

Shareholder value depends upon who the *shareholders* are. Not only on the capitalist *stockholders*, in the strict sense of the term, but also on employees, customers, citizens, and so many *stakeholders*. They should meet economic, social, environmental, and, yes, ethical practices.[24] They should abide by the precept that *"the only valid purpose of a firm is to create a customer"* that *Peter Drucker* recommended in 1973 in [62, *The Practice of Management*]. Maximizing only shareholder value may lower the welfare of so many other parties. Moreover, this maximization process is a short-term focus. By shrinking the temporal window during which the decisions are taken, short-term decisions can be detrimental to the rest of the society to the point of destroying the economy, let alone the ... long-term value that shareholders want to confiscate. There is no reason why a brief impulse of stock value, used as a unique criterion, is related to a productive long-term impact on this very same value.

We rather suggest to choose other criteria than shareholder value that speculators slowly succeeded to propagandize for their exclusive profit. *The duration of shareholders investments could be used as another criteria.* Or, a finer and more tractable one, the fluidity of the investment duration, which is its velocity.

The higher the fluidity of the investment duration, the shorter is the temporal window of investment in which the investment is made, the lower this fluidity, the larger its temporal window.

1. Slowing down those velocities;
2. And widening the aperture of the temporal windows (period of investments);

as a part of a regulation economic mechanism could be an objective of salubrious corporate management and sane (not insane) financial systems more and more governed by shareholders and less and less by the civil society and their elected representatives in the government. May be less corruptible, since their power depends upon the results of periodic elections. But the same financial power is used to corrupt also the election campaigns and procedures.

The question of taxing shareholder values is raised and could be studied mathematically. A Shareholder Value Tax could be added to a "Value Added Tax" on financial transactions as advocated in 1972 by *James Tobin*, which just confines the volatilities of financial transactions, consequences and not causes of investment durations too brief to be adequate for sound and productive investment and economic constraints, in our (debatable) opinion. Both transactions (velocity of commodities exchanged) and fluidities (velocity of investment durations) could and should be adapted to regulate viable economic evolutions, and thus taxed for doing so.

---

[24]Such as accounting scandals in 2001–2002 with Enron and so many others, the options backdating scandals of 2005–2006, and the subprime crisis of 2007–2008, which did not stop and even thrive after the crisis, up to the 2012 Libor Fraud.

1.5 The Ockham Razor

This study deals with the derivation of dynamic behavior of the agents and of the Market from the data of various types on *economic and financial constraints*. Instead of specifying dynamic "behavioral" transaction functions of agents (in a dynamical setting) or demand functions (in a static or intertemporal framework), we shall derive *only from the constraints and observation mechanisms* what transaction functions and pricing mechanisms can be for the constraints to be viable.

Viability theory designs and develops mathematical and algorithmic methods for investigating the *adaptation to viability constraints of evolutions governed by complex systems under uncertainty* that are found in many domains involving living beings, from biological evolution to economics, from environmental sciences to financial markets, from control theory and robotics to cognitive sciences (see [8, Neural Networks and Qualitative Physics: a Viability Approach]). It involves interdisciplinary investigations spanning fields that have traditionally developed in isolation.

One specific purpose of viability theory is to attempt to answer directly the question that some economists, biologists, or engineers ask: *"Complex organizations, systems and networks, yes, but for what purpose?"* The answer we suggest: *"to adapt to the environment."*

This is the case in economics when we have to adapt to real scarcity constraints, balances between supply and demand, and many other constraints, and in finance, to comply to budgetary rules, abstract, arbitrary, and ideological.

However, there is no reason why collective constraints are satisfied *at each instant* by evolutions under uncertainty governed by evolutionary systems. This leads us to the study of *how to correct* the lack of viability

1. either by changing the constraints, using for instance the *viability kernel* of the environment (see Definition 8.1.1, p. 119);
2. and/or by modifying the dynamics, using *viability multipliers* (see Sect. 12.1 of [15, *Viability Theory. New Directions*]),

in order to restore viability.

This may allow us to provide an explanation of the formation and the evolution of regulons (regulating controls) through regulation or adjustment laws that can be designed (and computed) to insure viability, as well as other procedures, such as using *impulses* (evolutions with infinite velocity) governed by other systems (see Sect. 12.2 of [15, *Viability Theory. New Directions*]), or by regulating the evolution of the environment (see [11, *Mutational and morphological analysis: tools for shape regulation and morphogenesis*] and [104, *Mutational Analysis. A Joint Framework for Cauchy Problems in and Beyond Vector Spaces*] by *Thomas Lorenz*).

Presented in such an evolutionary perspective, this approach of (complex) evolutionary systems departs from main stream modeling by a direct approach:

10 ***[Direct Approach]*** *It consists in studying properties of evolutions governed by an evolutionary system used as a* "model"*: gather the larger number of properties of evolutions starting from each initial state. It may be an information both costly and useless, since our human brains cannot handle simultaneously too many observations and concepts.*

Moreover, it may happen that evolutions starting from a given initial state satisfy properties which are lost by evolutions starting from another initial state, even close to it (sensitivity analysis) or governed by perturbed dynamical systems (stability analysis).

Viability theory rather uses instead an *inverse approach* for providing *mathematical metaphors*:

11 ***[Inverse Approach]*** *A set of prescribed properties of evolutions being given, study the (possibly empty) subsets of initial states from which*

1. *starts* ***at least*** *one evolution governed by the evolutionary system satisfying the prescribed properties, subset providing a qualitative evaluation of viable "contingent uncertainty";*
2. ***all*** *evolutions starting from it satisfy these prescribed properties, subset providing a qualitative evaluation of viable "tychastic" uncertainty.*

These two subsets coincide whenever the evolutionary system is deterministic.

We face the following dilemma: are the examples of financial and economic cycles produced by complex systems of differential equations with simple dynamics or produced by very *simple dynamics*, but confronted to a *complex maze of constraints*? We shall attempt to follow the second term of the above alternative (see Sect. 4, p. 57).

Numbers, next measurements, and then, statistics, how helpful they are for understanding and controlling the physical part of the environment, may be a drawback to address the qualitative aspects of our world, left to plain language for the quest of elucidation. We may have to return to the origins and explore new "qualitative" routes, without overusing the mathematics that our ancestors accumulated so far and bequeathed to us.

Meanwhile, we are left with this paradox: "simple" physical phenomena are explained by more and more sophisticated and abstract mathematics, whereas "complex" phenomena of living systems use, most of the time, relatively rudimentary mathematical tools.

For instance, the mathematical tools used so far did not answer the facts that, for instance,

1. economic evolution is *never* at equilibrium (stationary state);
2. and thus, there is no need that this evolution converges to it, in a stable or unstable way;
3. cognitive sciences does not recognize the overall rationality assumption of human brains;
4. and even more that they can be reduced to utility functions (numbers are not sufficient to grasp satisfaction);
5. that uncertainty can be mathematically captured only by probabilities (numbers, again);
6. that chaos, defined as a property of deterministic system, is not fit to represent a nondeterministic behavior of living systems which struggle to remain as stable (and thus, as "non-chaotic") as possible;
7. that intertemporal optimality, a creation of the human brain to explain some physical phenomena or what it has created himself, assumes that
    a. the existence of an actor (agent, decision-maker, controller, etc.),
    b. an optimality criterion,
    c. decisions are taken once and for all at the initial time,
    d. a knowledge of the future or of its forecasting (see Sect. 5.6, p. 80).
8. that evolving creations of Nature, are, for most of them, *myopic, exploratory but lazy, opportunistic but conservative*, and can at least be regulated, but not controlled by "decision-makers,"
    - *myopic*, since the knowledge or the future is not required;
    - *exploratory*, thanks to the contingent redundancy provided by the set of available regulons;
    - *lazy*, since, at each instant, (the norm of) the velocity of the regulon could be minimized.
    - *opportunistic*, since the regulatory controls (*regulons*, in short) have to be modified whenever viability is at stakes: *decisions should be taken at the right moment* (*kairos*, in classical Greek), *and need not to be optimal, which require an optimality criterion*;
    - *Conservative*, since the controls are kept constant whenever the viability constraints are satisfied.

    and that they need only to be viable, i.e., that their transient evolutions should be *adapted to their environments*;

among many other facts.

These observations led to dissent from the mathematical and economical mainstreams for forging some mathematical tools more or less adapted to overcome the shortcomings of the physical legacy. The economic validation of the mathematical assumptions and conclusions are the duty of economic and financial scholars and practitioners.

As often in economics or in life sciences, we face the following dilemma:

12 ***Simple Dynamics and Complex Constraints*** *Are the examples of economic evolutions produced by complex systems of differential equations with (or without) simple constraints or produced by very* simple dynamics, *confronted to a* complex maze of constraints*?*

Since it is easier to have access to constraints rather than to dynamics governing the complex behavior of agents, we shall use the Ockham law of parsimony:

13 ***Ockham's Razor*** Ockham's Razor *is the principle proposed by* William of Ockham *[1285–1347]:* "Pluralitas non est ponenda sine neccesitate"*, which translates as* "entities should not be multiplied unnecessarily"*. This "law of parsimony" states that an explanation of any phenomenon should make as few assumptions as possible, and to choose among competing theories the one that postulates the fewest concepts.*

These epistemological considerations are developed in Chap. 3 of [12, *La mort du devin, l'émergence du démiurge. Essai sur la contingence, la viabilité et l'inertie des systèmes*].[25]

These are some of the motivations to associate *duration functions* with chronological time, which flow with constant velocity and plays the role of a numéraire of evolutions. They are defined on fleeing temporal windows which evolve with them. These durations functions offer mathematical metaphors of a "specious time" passing more or less slowly. They are no longer prescribed, but *chosen* among available ones and *regulated*: we shall study in Chap. 7.1, p. 105 how the economic variables, the endowment function and the temporal windows on which the duration functions are defined *"co-evolve"* to guarantee the economic constraints.

1.6 Uconomia

How Did Economists Get It So Wrong?, asked *Paul Krugman* in an article published on September 6, 2009. *"As I see it, the economics profession went astray because economists, as a group, mistook beauty, clad in impressive-looking mathematics, for truth."* We cannot agree more with the dangers of the seduction of mathematics, in particular of the temptation to take theorems for reality. Economics, to be convinced

[25] *The Demise of the Seer, the Rise of the Demiurge. Essay on contingency, viability and inertia of systems*, waiting for a publisher who will take the same risks that Éditions Beauchesne took in publishing this 900 page book fitting no standard categories for translating it in English.

to become a true "science" in the sense that it deals with measurable quantities[4–11], got rid of the adjective "political" formerly qualifying economics, opening the gap between economic and social sciences, guilty of "mathematicism," as one speaks of "scientism." In the same way that utopia and uchronia have been invented, the economy described by the human brain is more of a u-economy, or, to avoid the unfortunate alliteration, an *uconomia.*[26]

However, are mathematics guilty of the charge: *"[...] During the golden years, financial economists came to believe that markets were inherently stable—indeed, that stocks and other assets were always priced just right. There was nothing in the prevailing models suggesting the possibility of the kind of collapse that happened last year."*?

This serious question deserves an elaborate answer.

Mathematicians provide theorems, proving that assumptions imply conclusions, or that assumptions are the price to pay *to buy conclusions*. If this price of the assumption is too cheap to be relevant, we obtain junk conclusions as we speak of junk bounds: they are not profitable, and may be counterproductive.

When they are used as mathematical metaphors in other disciplines, translation of mathematical concepts have to be *validated*[27] by their specialists. This is another important question, because the "translation" of theorems true in mathematics may induce statements which are not consistent in the framework of the discipline. If the translated conclusion is "false," this implies either that some assumptions are not satisfied or that the metaphor is not validated.

Taking advantage of inescapably esoteric mathematics, requiring a forbidding access time to allow us their manipulation, those metaphors can be easily hijacked from their mathematical context to justify this or that ideology. The less validated, the more these metaphors are mathematically adorned.

For instance, *static economics* used the basic tools of optimization and equilibrium theorems for explaining the exchange value and the emergence of prices: taking into account scarcity constraints on commodities, prices, and monetary values, it *assumes the Walras law: it is forbidden to spend more in monetary values than what is received in an instantaneous transaction*. In other words, the value of the transaction must be negative or equal to zero.

This is indeed a *sufficient* condition (by no means necessary) for proving a mathematical theorem on the existence of a Walrasian (general) equilibrium. Knowing

1. the behavior of the agents described by their *demand functions* of commodities depending on prices and incomes;
2. the set of available commodities to allocate among agents;

[26]See Sect. 5.2 of [12, *La mort du devin, l'émergence du démiurge*].

[27]Physics use experimentations to establish this validation task. Life sciences seldom allows the possibility of experimentation or because the chosen variables have no units of measure.

3. and an *arbitrary feasible allocation of a collective income* derived from the value of the set of available commodity at each price,[28]

there exists an "equilibrium price." This is at this point that the concept of equilibrium became polysemous. In mathematics and physics, as well as for *Léon Walras*, an equilibrium is a stationary state of a dynamical system, as the Walras tâtonnement process in economics. Starting from an equilibrium, the system does not evolve. Otherwise, prices are *assumed* to evolve according to the Walras tâtonnement, the so-called supply and demand law, *forbidding transactions before the infinite time when the allocation of commodities is* viable *in the sense that the sum of allocated commodities belongs to the set of available commodities.*[29] Non Walras tâtonnement systems, or, rather, bilateral tâtonnement processes governing both the evolution of commodities and prices have been studied by many authors.[30]

Note that it is this very *"viability property"* which does matter, not the fact that this price is an equilibrium in the mathematical sense. Hence *disequilibrium* means the violation of either one of these two meanings. Since we assume that the economic constraints depend on time, there is no (stationary) equilibrium anyway (keeping one component constant, a price, for instance, leads to some concept of disequilibrium, for instance). Disequilibrium thus means in this case *the violation of viability*. We rather use the term *disviability* (or non-clearing markets) instead of using the polysemous concept of "disequilibrium.[31]" Therefore the standard Walrasian tâtonnement is disviable at every finite time and viable only when ... time becomes infinite, except if we start from a Walras equilibrium and remain there forever!

The Walras law is used as an ideological law and needs to be specified: *who allocates the incomes (monetary endowments) among the agents?* The answer is the "invisible hand" of Adam Smith, the undefined "Market" or a "Market Secretary," who, then, fabricates an "Invisible Gosplan."

The static answer *assumes* that the endowment of the agent is derived from the commodity owned before the transaction, so that the income of the agent is the value of the endowed commodity. The values of the transactions must remain nonpositive in this framework. Consequently, *money, used for allocating incomes to the agents,* ***evaporates*** *from the static Walrasian framework.* To go from statics to dynamics in the absence of money requires that, at each time, *the price is given at each date*

[28]This assumption is rarely mentioned by the proponents of the *fair efficiency* of the "Marker." Allocating $\varepsilon\%$ of the total income to $n-1$ agents and $(1-(n-1)\varepsilon)\%$ to the remaining one leads to a Walras equilibrium.

[29]See for instance [55, Debreu], [4, Arrow & Hahn], [88, Hildenbrand W. & Kirman], [5, Aubin], [34, Bewley], [119, Stachurski, Venditti & Yano], etc.

[30]Among them, [83, Hahn & Negishi], [112, Ostroy & Starr], [116, Smale], [121, Uzawa], etc.

[31]See [118, Solow & Stiglitz], [27, Barro & Grossman], [61, Drèze G. & de la Vallée Poussin], [47, 48, Champsaur, Drèze & Henry], [111, Neary & Stiglitz], [72, Gale], [30, 31, Bénassy] [56, De Vroey] for instance.

by the static Walras allocation mechanism. For that purpose, the static Walras law should be obeyed at each time.

This is a nightmare that can kill an economy and, actually, did destroy some of them in many historical examples.

Why should this static Walras law be required, whereas, *an evolutionary economy involves transactions*? Whenever time is involved, there is no reason to abide by this static Walras law. *Credit* has been slowly invented *for transferring in the future abstract positive transaction values* for allowing present transactions of real commodities. In other words, credit allows the agents to obtain commodities which, being available at present time, may not be affordable by lack of monetary units to finance the transaction. The question arises to understand from where should come the credit, this necessary monetary endowment lubricating evolving transactions.

Consequently, knowing

1. the behavior of the agents described by their *transaction functions* (and no longer their demand functions) depending on commodities and prices;
2. the set of available commodities to allocate among agents,

compute at each instant the amount of money endowed to the agents for

1. allowing the agents to realize their transactions;
2. helping the invisible "Market" to regulate the prices (see [9, *Dynamic Economic Theory: A Viability Approach*]).

Accordingly, the Adam Smith's invisible hand does belong to an *invisible man* who should use his other hand to advise the agents to perform their transactions.

Still invisible, "He" could then pilot *both* the prices of the Market with one hand and the transactions of the agents with the other one, in a consistent way *for the benefit of the agents*, not necessarily for the "Market."

Therefore, *for providing agents viable commodities*, which could be a reasonable objective, we need to study what a viable endowment of credit policy could be, allowed to violate the sanctified *static* Walras law during some adequate temporal windows if needed, viability oblige. In an evolutionary framework of economic variables made of transactions of commodities and prices, the aggregated values of the transactions fluctuate. This is what we observe, between financial crises, such as the ones of 1929 and 2008, caused by inadequate and *deregulated* monetary endowments, too small for guaranteeing economic viability, too large for triggering financial exuberance. Monetary endowments, even as abstract numbers hidden in computers, are crucial because of their economic consequences, unemployment, poverty, etc.

Therefore, we should try as much as we can to be agnosticist and not to abide by the "banker faith" in general equilibrium theory, as blind as the usual one, *pretending that it is supported by scientific reasoning or mathematical proofs*. This is not really the case since the available mathematics, motivated by the physical sciences, are not yet designed to account for the phenomena of the living world, to which our behaviors belong (see [12, *La mort du devin, l'émergence du démiurge*]).

However, since the present-day economic *doxa* is based on general equilibrium theory, fortified by so many mathematical theorems based on optimization and nonlinear analysis,[32] the evolutionary economic motivations described in the introduction are also based on other mathematical results, based on set-valued analysis, differential inclusion, and viability theory.

The doxa dealt with the congelation of the admirable Walras tâtonnement and its associated equilibrium. At that time, the end of nineteenth century, our contemporaneous mathematical tools were not available. One cannot reproach to the pioneers such as *François Quesnay*, *Victor Mirabeau*, *Turgot*, d'*Alembert*, *Borda*, *Condorcet*, *Nicolas Canard*, *Augustin Cournot*, *Auguste Walras*, his son *Léon Walras* (a socialist economist, according to *Charles Péguy*), *Vilfredo Pareto*, etc., to have dared using a mathematical approach for grasping social and economic issues despite great opposition. Later, *Luitzen Brouwer*, *John von Neumann*, *Shizuo Kakutani*, *John Nash*, *Kenneth Arrow*, *Gérard Debreu*, *Robert Aumann*, *Lloyd Shapley*, *Herbert Scarf*, *Ky Fan* and many others contributed to forge and develop difficult theorems on equilibria and optima justifying the concepts of static economies. The concept of time and evolution was not forgotten, however: it was involved in the works of *Alfred Marshall*, *John Hicks*, *Friedrich Hayek* and *John Maynard Keynes*, among too many others to quote them.

The danger of accepting the static equilibrium point of view was recognized in 1944 by *John von Neumann* and *Oscar Morgenstern* at the end of the first Chapter of their monograph *Theory of Games and Economic Behavior*: *"Our theory is thoroughly static. A dynamic theory would unquestionably be more complete and therefore, preferable [...]"* It was time to go from static to genuine evolutionary systems, which are not to be confused with the concepts of "repeated equilibria," which required that at each instant, an *instantaneous equilibrium mechanism* provides the prices and the commodities obtained through a *demand function* depending on price and income.

We single out a simpler and more appealing concept of *(instantaneous) transaction*, which is the *velocity of an evolving commodity*, and suggest to replace demand functions by *transactions functions*, depending on the commodities, the prices and the monetary endowments.

To keep monetary endowment always equal to 0 became the Grail of many economists, bankers, and politicians, instead of computing it at each time for guaranteeing the viability of the economic system.

What may be wrong with present day orthodoxy deals not only with social and economic issues, but also with mathematical theories supporting them. *Mathematics evolve*, too, increasing the tool-box offered to users, even though if most of them were motivated by physical and engineering sciences and not by social and economic ones.

---

[32]See for instance [5, *Mathematical Methods of Game and Economic Theory*] and [10, *Optima and Equilibria*] to show that we, too, were seduced, not for long, by general equilibrium theory.

Economics is still identified with the exclusive and exclusively use of equilibrium and optimization theories as chance in finance is automatically mathematically translated by stochastic differential equations, as if there were still the only mathematical approaches to be offered.

This may be what *Paul Krugman* had in mind.

Questioning the mathematical economics *doxa* on the mathematical side is the purpose of this study, hopefully strengthening heterodox proposal in economics, reacting to the pernicious features of general equilibrium theory, narrowly identified with mathematical economics.

This does not guarantee that the economic motivations are relevant and that the conclusions provide better explanations and guidelines, but, at least, *they cannot be dismissed under the pretext of a lack of mathematical rigor.*

This is the reason why some proofs are provided, and references made to Chaps. 13, p. 523, 15, p. 603 and 17, p. 681, of [15, *Viability Theory. New Directions*]. What is important is the translation of mathematical language to economically relevant issues, hoping that economists will debate our conclusions and *question our assumptions*, with no need to be involved in the intricacies of mathematics that they did not have the opportunity to become familiar with. It is enough to trust the mathematicians to testify that this part of the study is technically sound.

Abandoning the identification of (mathematical) economy with equilibrium theory and a pervading optimization behavior of economic agents and trading it for a dynamical behavior of economic agents in search of viable instead of optimal evolutions has been a view rejected by most academic economists. Ever since the first study [6, *A dynamical, pure exchange economy with feedback pricing*] presented in a high standard economic journal in 1981, for this very same reason: it was not economically relevant since there was no utility function nor equilibrium. This is still the current argument more than 30 years later for ignoring studies along these lines. Very recently (2012), an editor of a famous review rejected without referee an article dealing with the topic of this book: *"I do not understand what role the economic agents play in the author's model. They are obviously not maximizing, but it* appears *that they also do not follow simple behavioral rules."* "Appearance" seems nowadays the scientific yardstick to hide the fact that no effort is made to read even the introduction where this behavior is described. Furthermore, *"the paper is kept at a highly* abstract *level"*, as if the "long march to abstraction" was not the main successful enterprize of mankind, whether good or bad its consequence!

In this short text, right or wrong, we do replace maximizing behavior by a dynamical one, guaranteeing viability instead of optimality, advising to choose decisions at the right moment rather than optimal[33] ones, and, for that purpose, following the route of increasing abstraction.

[33]Under which criterium or subset of criteria, most of the time alluded to?

The "inertia principle" (see Definition 21, 58) and the "Ideal Heresy Law"[34] explain this behavior.

This has always happened, as the history of science and technology teaches us. Crazes and irrational fads rule parts of academic science as many other human activities: *Charles Mackay*, author of the 1841 *Memoirs of Extraordinary Popular Delusions and the Madness of Crowds* on bubbles or financial manias among many other Panurgean or mimetic behaviors, summarized it in his poem:

You have no enemies, you say?
Alas, my friend, the boast is poor,
He who has mingled in the fray
Of duty, that the brave endure,
Must have made foes.
If you have none,
Small is the work that you have done.

[34] *The* "Ideal Heresy Law", *like the* "Ideal Gas Law", *states that the intensity of the struggle against heresy is inversely proportional to the deviance that separates it from dogma. The enthusiasm of new converts to transform an old heresy new dogma is proportional to the violence they practiced for the persecuted before denying it. New beliefs*—credulitas—*are established mostly at the cost of thousands cruelties*—crudelitas (see Box 173, Sect. 4.4.5, of [12, *La mort du devin, l'émergence du démiurge*]).

Chapter 2
How Long and How Much Endowing One Commodity

Real economic problems are complex and involve too many variables and constraints to be both computable and reasonably relevant. Mathematics cannot be "applied" to "economics," at best, economics can motivate mathematics, in other ways than physical sciences did, by offering at least "qualitative" mathematical metaphors of economic evolutions, not quantitative ones[4–11]. But the access time to these statements is forbidding. Consequently, we suggest to begin our study by "illustrating" the general concepts in the context of the simplest example of the evolution of a single commodity. These illustrations are quite instructive, provide another way than general mathematical theorems to grasp what the concepts attempt to mean and to assess their economic relevance. However, in any case these simple systems are far too simplistic for their conclusions to be taken seriously. But if a given property is not economically validated for such an example, it is "falsified" in the Popperian sense in the general case.

Since we deal with evolutionary economic systems, we need to start with some considerations on time and evolution, on prediction and anticipation, by restricting the polysemy of these concepts as much as we can.

2.1 Provider Financing One Commodity

We study here the evolution of (only) one commodity $t \mapsto x(t) \in \mathbb{R}$ (for instance, the number of units of a numéraire).

Assume that $\mathbf{k} \in \mathbb{R}$ is a *minimal threshold* (or, in this economic setting, a *viability threshold*.[1]) above which the commodity must remain.

[1]In which case $\mathbf{k} \geq 0$.

J.-P. Aubin, *Time and Money*, Lecture Notes in Economics and Mathematical Systems 670, DOI 10.1007/978-3-319-00005-3_2,

The simplest example of viability condition requires that there exists a propitious temporal window $[T - \Omega, T]$ of *aperture* Ω such that the *viability property*

$$\forall\, t \in [T - \Omega, T], \;\; x(t) \;\geq\; \mathbf{k} \tag{2.1}$$

holds true. This provides in particular the *initial value* $x(T - \Omega)$ at the beginning of the temporal window whenever any terminal condition $x(T) = x$ is required at its end.

In economics and finance, the velocity[2] $x'(t) = \lim_{h \mapsto 0+} \dfrac{x(t) - x(t-h)}{h}$ is interpreted as a *transaction*.[3] In this chapter, we require that the transaction is equal to the *endowment* needed to finance it. The coevolution of the commodity and of the endowment is described by the system of two differential equations[4]

$$\begin{cases} (i) \; x'(t) \;=\; y(t) \\ (ii) \; y'(t) \;=\; \varrho(y(t)) \end{cases} \tag{2.2}$$

satisfying the terminal condition

$$x(T) \;=\; x \tag{2.3}$$

where $x \geq \mathbf{k}$ is a parameter denoting the objective to be reached at time T.

In this illustration of otherwise general concepts avoiding unnecessary cumbersome analytical details, we assume that $\varrho \;=\; 0$, so that the variables y denote *constant endowments*, which vary as physical parameters do in deterministic dynamical systems: they remain constant on the temporal widow $[T - \Omega, T]$ and wait for the evolution to end at the temporal window before being changed.[5]

---

[2]According to a suggestion of *Efim Galperin*, derivatives from the right $x'(t) = \lim_{h \mapsto 0+} \dfrac{x(t+h) - x(t)}{h}$ are "physically non-existent" since time $t + h$ is not yet known, following the Czech Count Jiri Buquoy, who in 1812, formulated the equation of motion of a body with variable mass, which retained only the attention of *Poisson* before being forgotten.

[3]If we regard the number $x(t)$ of units of numéraire as a *value*, then $x'(t)$ is regarded as an *impetus*, which is the derivative of a value, in the same way than in mechanics, the force is the derivative of a potential, the analog of value.

[4]Stating that the acceleration $x''(t) = \rho(x'(t))$.

[5]This issue, at the origin of the gap between specialists of dynamical systems and those on control theory (where the parameters evolve), is rarely explicitly dealt with. This is the reason for mentioning the following quotation of the book [127, *Introduction to Applied Nonlinear Systems and Chaos*] by *Stephen Wiggins*: ***On the Interpretation and Application of Bifurcation Diagrams: A Word of Caution***: "*[...] In all of our analysis thus far the parameters have been constant. The point is that we cannot think of the parameter as varying in time, even though this is what happens in practice. Dynamical systems having parameters that change in time (no matter how slowly!) and that pass through bifurcation values often exhibit behavior that is very different from the analogous situation where the parameters are constant.*"

Recall that a *set-valued map* $F : X \rightsquigarrow Y$ associates with any $x \in X$ a subset $F(x) \subset Y$. This subset $F(x)$ can be empty, reduced to a singleton $F(x) := \{y\}$ or equal to a subset containing more than one element. It is said to be *single-valued* if for any $x \in X$, $F(x) := \{f(x)\}$ contains only one element $f(x)$.

Definition 2.1. [Regulation Maps] A *regulation map* $R : (t,x) \rightsquigarrow R(t,x)$ (regarded as a *transaction map* in economics and a *management rule* in finance) associated with the system of differential equations (2.2), p. 32, is a map such that the evolution governed by the *regulation law*

$$\forall\, t \in [T - \Omega, T], \;\; x'(t) \;=\; R(t, x(t)) \tag{2.4}$$

is also governed by the system (2.2), p. 32, but satisfies additional requirements (such as viability and other ones).

Happily, it may happen that either $R(t,x)$ is single-valued or that $R(t,x) = R(x)$ does not depend on time, or both: in this case, the regulation map is called a *feedback* (in automatics).

1. In the framework of a *direct approach*[10–21], the regulation map R is given as a part of the *assumptions* of the "model" (and most often, assumed to be single-valued).

   For instance, we could just assume that, as *Thomas Malthus* did in his celebrated *An essay on the principle of population* (1798) in the framework of population dynamics,[6] that $x'(t) = -rx(t)$, providing exponential $x(t) = e^{-r(T-t)}x$ which violate the viability constraint when $\mathbf{k} > 0$: it is viable on the temporal window $\left[T - \frac{\log\left(\frac{x}{\mathbf{k}}\right)}{r}, T\right]$.

   Or, to overcome this sad property, to follow *Pierre-François Verhulst* who proposed in 1838 the *"logistic"* equation $x'(t) = -rx(t)(x(t) - \mathbf{k})$. It provides the *logistic* (or "sigmoid") function $x(t) = \frac{\mathbf{k}x}{x + (\mathbf{k} - x)e^{-r(T-t)}}$, arriving at x at time T, remaining above $\mathbf{k}$ on the temporal window $]-\infty, T]$, decreasing and converging to it asymptotically.

   Or the thousands macro-economic models a priori cleverly designed, but quite often not viable.

[6]The sign of the growth rate is negative instead of positive, since we want to bound below the number of commodity units instead of bounding above the members of the population: *"Population, when unchecked, increases in a geometrical ratio"* and his solution: *"By moral restraint, I mean a restraint from marriage [...]"*. In economics, this would be *a restraint from consumption*. See Chaps. 6 and 7 of [15, Aubin, Bayen & Saint-Pierre] for more details.

2. Viability is one of the properties required in the framework of an *inverse approach*[11–21]: we are interested by the *construction* and the *computation* of *regulation maps* $(t,x) \rightsquigarrow R(t,x)$ and of the aperture $\Omega \geq 0$ of the temporal window $[T-\Omega, T]$ over which the evolution governed by $x'(t) \in R(t,x(t))$ guaranteeing at least the viability property (2.1), p. 32 and other ones, such as the ones listed below.

2.2 Restoring Viability

Before mentioning them, just few obvious observations are in order. The evolutions governed by (2.2), p. 32 are affine functions

$$\forall\, t \in \mathbb{R}, \;\; x(t) \; = \; x - y(T-t) \tag{2.5}$$

where y is a constant endowment (velocity) and x the terminal condition at T. It is viable on the temporal window of aperture $\Omega := \dfrac{(x-\mathbf{k})}{y}$.

Either the endowment $y = x'(T-\Omega^{\flat}) < 0$ is negative, and thus the scarcity requirement is violated, or $y^{\star} = x'(T-\Omega^{\flat}) \geq 0$, and the evolution remains viable at least for some time. If $y = x'(T-\Omega^{\flat}) = 0$, then $x(t) = \mathbf{k}$ for all $t \in [T-\Omega^{\flat}, T]$, and fails to satisfy the terminal condition except if $x = \mathbf{k}$. If $y > 0$, then the solution reaches x at time T if $y = \dfrac{x-\mathbf{k}}{\Omega^{\flat}}$ where $\Omega^{\flat} := \Omega^{\flat}(T,x,y) := \dfrac{x-\mathbf{k}}{y}$ is the aperture of the temporal window $[T-\Omega^{\flat}(T,x,y), T]$ on which x can be reached at time T by the evolution starting from $\mathbf{k}$ at time $T-\Omega^{\flat}(T,x,y)$.

Hence, as soon as an evolution hits the viability threshold at time $T - \Omega^{\flat}(T,x,y) := \frac{x-\mathbf{k}}{y}$, the terminal state can be reached with the endowment y satisfying $\dfrac{x-\mathbf{k}}{\Omega^{\flat}(T,x,y)}$.

Observe that *starting from any initial state* $x_{T-\Omega} > \mathbf{k}$ *at the beginning* $T-\Omega$ *of a temporal window* $[T-\Omega, T]$, *any evolution is viable* either *on* $[T-\Omega, T]$ *until it reaches* x *at time* T or *there exists some aperture* $\Omega^{\flat} \in [0, \Omega]$ *when* $x(T-\Omega^{\flat}) = \mathbf{k}$ *reaches the viability threshold.*

This is the time when the viability question arises for providing a new (positive) endowment or/and a new commodity (strictly larger than the viability threshold) for allowing the evolution to restart and reach x at time T:

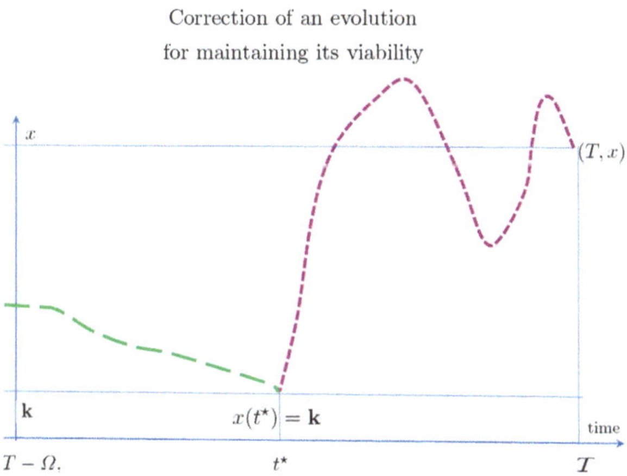

14 ***[Restoring Viability]*** *Whenever viability is at stakes at the time* $t^\sharp = T - \Omega^\flat$ *in the sense that* $x(t^\sharp) = \mathbf{k}$*, how can we choose an endowment* $y^\star = x'(t^\sharp)$ *or/and a commodity* $x^\star > \mathbf{k}$ *such that the solution* $x(\cdot)$ *governed by the system (2.2), p. 32 starting at* $(t^\sharp, \mathbf{k})$ *with a velocity* $y^\star$ *is viable on* $[T - \Omega^\flat, T]$ *and reaches* x *at time* T*?*

Any requirement added to answer this question will define an adequate regulation map R allowing to restore the viability by evolutions governed by differential inclusion $x'(t) \in R(t, x(t))$ on the temporal window $[T - \Omega^\flat, T]$.

We provide a short list of examples of additional requirements.

1. **Required Aperture.** The aperture Ω is given and the viability is required on the whole temporal window $[T - \Omega, T]$, i.e., $\omega^\star(T, x, y) = \Omega$. Then the regulation map R_Ω is obviously defined by

$$y \;=\; R_\Omega(T, x) \;:=\; \frac{x - \mathbf{k}}{\Omega}$$

so that we have computed this feedback in this simple case, as well as the initial commodity $x(T - \Omega) = \Omega R_\Omega(T, x)$.
2. **Viability Threshold Endowment Condition.** When the evolution $x(\cdot)$ hits the viability threshold $\mathbf{k}$ at time $t^\sharp$, the endowment is fixed as a function $\Delta(t^\sharp)$. The function $t \mapsto \Delta(t)$ is called a *viability threshold condition*. This viability threshold condition being given, we construct a regulation map $R_\Delta(T, x)$ (which may be set-valued when the viability threshold condition is not monotone) providing the set of endowments provided when scarcity is at stakes for reaching x at time T (see Sect. 2.3, p. 37).
3. **Viability Threshold Impulse Condition.** Instead of imposing the endowment through a viability threshold condition when the evolution $x(\cdot)$ hits the viability

threshold $\mathbf{k}$ at time $t^\sharp$, we allow the evolution of the commodity to be discontinuous and to "jump" at a new level $x^+ = \Phi(t^\sharp)$ with an *impulse* (infinite velocity). Starting from this new condition $x(t^\sharp) = \Phi(t^\sharp)$, we build a regulation map R_Φ such that any velocity $y \in R_\Phi(T, x)$ will allow to reach x at time T. Such a system is an example of *impulse dynamical systems*. See more details in Sect. 12.2 of [15, Aubin, Bayen & Saint-Pierre] and in the recent book [74, *Hybrid Dynamical Systems*] by *Rafal Goebel* et al.

We can combine the viability threshold condition and impulse requirements by associating with time $t^\sharp$ the new state $x(t^\sharp) = \Phi(t^\sharp)$ and the velocity $y(t^\sharp) = \Delta(t^\sharp)$ (see Sect. 2.5, p. 41).

4. **Governing the Evolution of Endowments.** For the time, we imposed that the endowments were constant parameters, governed by the differential equation $y'(t) = 0$. We may require that the velocity $y'(t)$ of the endowments depend linearly on the velocities $x'(t) = y(t)$, i.e., that the evolution of the endowments is from now on governed by a differential equation

$$y'(t) = \mathbf{m}(t)y(t) + \mathbf{l}(t) \tag{2.6}$$

reaching $\mathbf{C}$ at time T.

We also allow the viability threshold $\mathbf{k}(t)$ of the endowment to depend on time (see Chap. 3, p. 47).

5. **Choosing Initial Endowments in a Set.** Imposing a unique endowment y when the commodity reaches the viability threshold may be too strong a requirement. We may bring a little more flexibility by requiring only that the endowment y can be chosen and evolve in a given interval $[0, c]$. In other words, we impose only a constraint on the endowment.

This flexibility is translated as (simple) differential inclusion

$$\begin{cases} (i) \ |x'(t)| = y(t) \\ (ii) \ y'(t) = 0 \end{cases} \tag{2.7}$$

See Chap. 4, p. 57.

6. **Endowment Depending on the Commodities and Prices.** Until now, we assumed that the endowment financed the transaction of one commodity under the assumption that the price was constant (and equal to 1). If not, prices $t \mapsto p(t)$ evolve and their velocities are regarded as *price fluctuations*.

We denote by

$$\begin{cases} U(x(t), p(t)) := p(t)x(t) \text{ the } \textit{patrimonial value} \\ E(x(t), p(t)) = p(t)x'(t) + p'(t)x(t) \text{ its } \textit{impetus} \end{cases}$$

We associate with an economic pair (x, p) and an endowment y the set $\mathcal{U}(x, p; y)$ of transaction-fluctuations (υ, π) defined by

$$\mathcal{U}(x, p; y) := \{(\upsilon, \pi) \text{ such that } p\upsilon + \pi x \leq y\}$$

Hence the coevolution of commodities, prices, and endowment is governed by the *differential inclusion*

$$\forall\, t \in [T-\Omega, T],\ \ (x'(t), p'(t)) \in \mathcal{U}(x(t), p(t), y(t)) \text{ and } y'(t) = 0 \qquad (2.8)$$

Therefore, concerning the financial aspect, the retroactions laws associating endowments $y \in R(t, \upsilon)$ with time t and patrimonial value υ that have been mentioned can be used to define the regulation maps $(x'(t), p'(t)) \in \mathcal{U}(x(t), p(t), R(t, p(t)x(t)))$.

Naturally, the situation is much more complex, because commodities $x(t) \in \mathbb{R}^\ell$ and prices $p(t) \in \mathbb{R}^{\ell\star}$ range over ℓ-dimensional spaces, or even, in the case of n economic agents, commodity allocations $x(t) = (x_1(t), \ldots, x_n(t)) \in \mathbb{R}^{n\ell}$ over a $n\ell$ dimensional one. Additional economic constraints of the form $(x(t), p(t)) \in K(t)$ may be added to the financial ones bearing on the impetuses

$$E(x(t), p(t)) := \sum_{i=1}^{n} \langle p(t), x_i'(t)\rangle + \langle p'(t), x_i(t)\rangle = \frac{d}{dt}U(x(t), p(t))$$

of the patrimonial value

$$U(x(t), p(t)) := \sum_{i=1}^{n} \langle p(t), x_i(t)\rangle$$

These issues are the topics studied in Chap. 7, p. 105.

We investigate in the framework of system (2.2), p. 32, when the endowment provider acts when the commodity reaches the *viability threshold* for financing the economic agent by providing him either some endowment or some amount of commodity above the threshold or both.

We begin by the case when it provides only an endowment and next, endowment and commodity.

2.3 Providing Endowment

Proposition 2.3.1. ***[The Threshold Endowment Regulation Map]*** *The viability threshold endowment function Δ being given, the associated regulation map R_Δ provides the aperture $\Omega^\flat \geq 0$ and the amount y of the endowment necessary to reach x at time T in a viable way. The apertures $\Omega^\flat$ are the fixed points of*

$$\Omega^\flat \Delta(T - \Omega^\flat) = x - \mathbf{k} \qquad (2.9)$$

and the amounts $y = \Delta(T - \Omega^\flat)$ of endowments is the subset $R_\Delta(T, x)$ of fixed points y of

$$y = \Delta\left(T - \frac{(x-\mathbf{k})}{y}\right) \tag{2.10}$$

The regulation map R_Δ is single valued if and only if the viability threshold endowment condition Δ is strictly decreasing.
Otherwise, whenever a pair of initial instants t_i satisfies $u(t_1,t_2) := \frac{t_1-t_2}{\Delta(t^1)-\Delta(t_2)} \geq 0$, then there exists a shock at time $t^\star = t_1 + u(t_1,t_2)\Delta(t_2) = t_2 + u(t_1,t_2)\Delta(t_1)$ and $x^\star = \mathbf{k} + u(t_1,t_2)y_1 = \mathbf{k} + u(t_1,t_2)y_2$.

Proof.

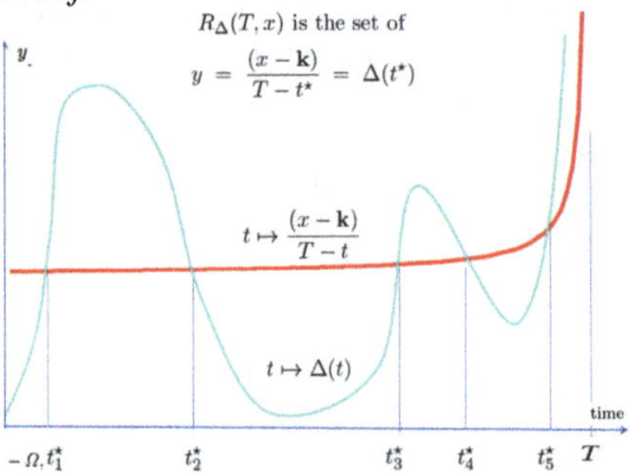

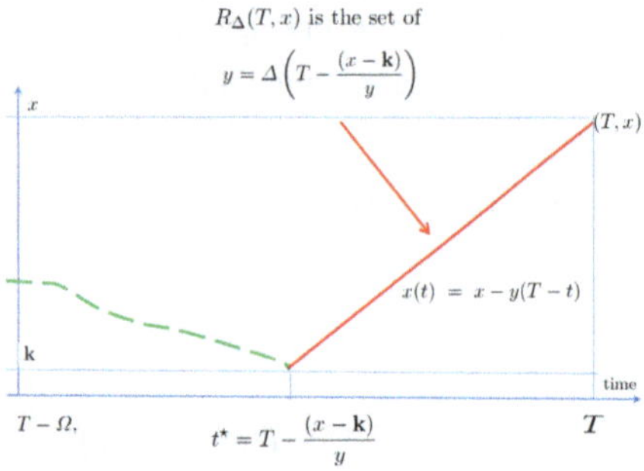

Let (T,x) be fixed and consider the evolution $x(t) := x - y(T-t)$ associated with the endowment y reaching the target $x(t^\sharp) = \mathbf{k}$ at time $t^\sharp := T - \Omega^\flat$ when

$$\Omega^\flat = \frac{(x-\mathbf{k})}{y} \geq 0 \text{ or } y = \frac{(x-\mathbf{k})}{\Omega^\flat}$$

Since the threshold condition requires that the endowment $y = \Delta(T-\Omega^\flat)$ is equal to $\Delta(T-\Omega^\flat)$, we infer that the aperture is a solution to the equation $\Omega^\flat\Delta(T-\Omega^\flat) = x - \mathbf{k}$ or, equivalently, that the endowment is a fixed point $y = \Delta\left(T - \frac{(x-\mathbf{k})}{y}\right)$.

Hence the regulation map $R_\Delta(T,x)$ is a subset, which can be empty (no solution), reduced to one element $\{r_\Delta(T,x)\}$ or the subset of values $y = \Delta(t^\sharp)$ such that $\frac{(x-\mathbf{k})}{T-t^\sharp} = \Delta(t^\sharp)$ containing more than one element.

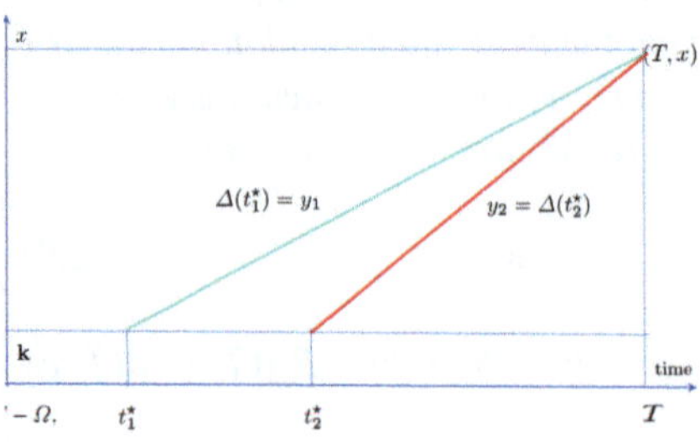

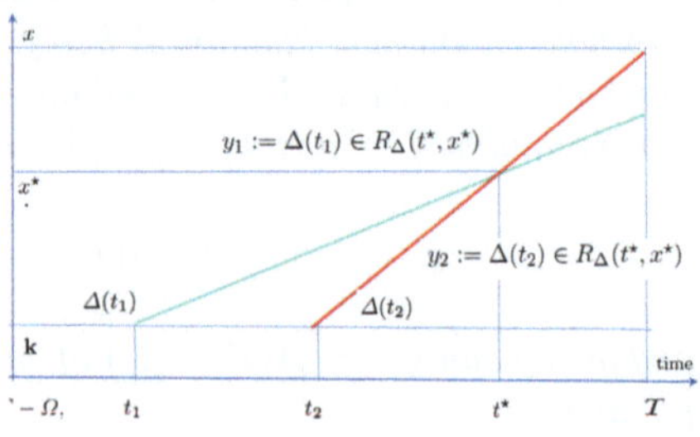

Assume that there exist two endowments $y_i \in R_{\Delta}(T,x)$, $i = 1,2$, such $y_1 < y_2$ and $t_i^\sharp := T - \frac{(x-\mathbf{k})}{y_i}$. Then

$$t_1^\sharp \; < \; t_2^\sharp \text{ and } \Delta(t_1^\sharp) = y_1 < y_2 = \Delta(t_2^\sharp)$$

1. If Δ is strictly decreasing, we would deduce that $\Delta(t_2^\sharp) \leq \Delta(t_1^\sharp) < \Delta(t_2^\sharp)$, which is impossible. In this case, the regulation map $R_{\Delta}(T,x) = \{r_{\Delta}(T,x)\}$ is single-valued (there is no shock).
2. Conversely, assume that Δ is "not strictly decreasing" in the sense than there exist $t_i \leq T$, $i = 1,2$, such that

$$t_1 \; < \; t_2 \text{ and } \Delta(t_1) \; \leq \; \Delta(t_2)$$

Let us consider the "characteristics" $x_i(t) := \mathbf{k} + \Delta(t_i)(t - t_i)$ starting at $\mathbf{k}$ at time t_i. They intersect at time $t^\star$ satisfying

$$t^\star \; = \; t_2 + y_1 \frac{t_2 - t_1}{y_2 - y_1} \; \geq \; t_2 \; \geq \; t_1$$

Denoting by $x^\star = x_1(t^\star) = x_2(t^\star)$ the common value of the characteristics, we infer that both $y_i := \Delta(t_i)$ belong to $R_{\Delta}(t^\star, x^\star)$, i.e., that a shock happens at time $t^\star \geq t_2$ and $x^\star$. ■

In the extent that set-valued maps are rejected for lack of economic (or physical) sense, we can provide selections of the map R_{Δ}, which are single-valued, but discontinuous. Among them, we single-out upper $r_{\Delta}^\sharp(t,x)$ and the lower selections $r_{\Delta}^\flat(t,x)$ defined by

$$\begin{cases} r_{\Delta}^\sharp(t,x) =: \max R_{\Delta}(t,x) \\ r_{\Delta}^\flat(t,x) =: \min R_{\Delta}(t,x) \end{cases} \tag{2.11}$$

Hence $R_{\Delta}(t,x) \;\subset\; [r_{\Delta}^\sharp(t,x), r_{\Delta}^\flat(t,x)]$. We deduce that the apertures satisfy inequality $\Omega^\flat(T,x,r_{\Delta}^\sharp(t,x)) \leq \Omega^\flat(T,x,r_{\Delta}^\flat(t,x))$.

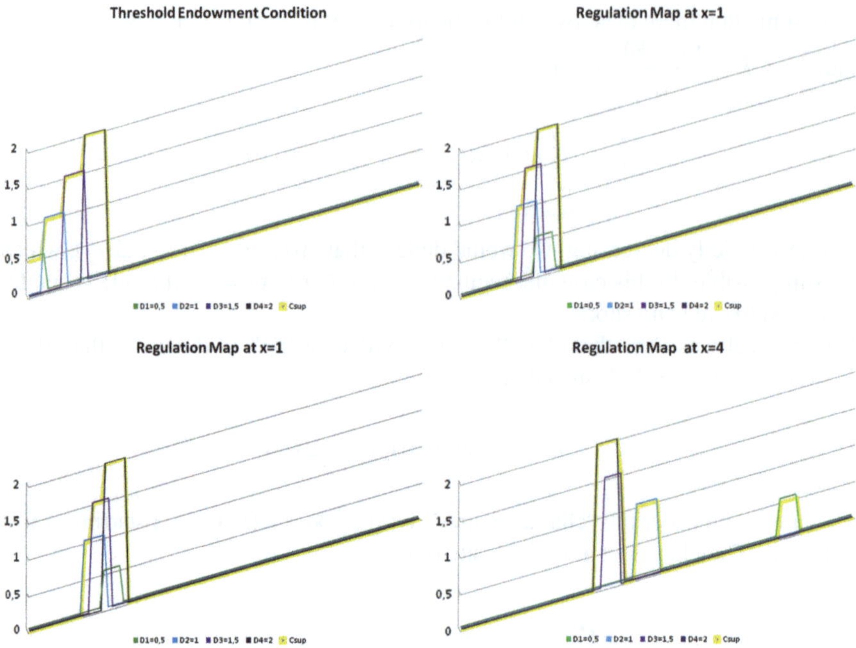

15 ***Regulation Map with Shocks.*** *Consider a viability threshold endowment function Δ which is a not strictly increasing* step *(or* staircase, piecewise constant*) functions with four steps, displayed in the left figure of the first row.*
For a fixed commodity x, the figure in the right displays the graph of $t \rightsquigarrow R_\Delta(t,x)$ which is set-valued map (having shocks*) as well as the graphs of its upper selection $r_\Delta^\sharp(t,x)$. Since the endowment condition is increasing, shocks may happen. Indeed, the three other figures display the graphs of the maps $t \mapsto R_\Delta(t,x)$ for three other values of the commodity, where shocks are produced.*

2.4 Viability Property of the Regulation Map

Lemma 2.4.1. *[**Viability Property**] Let $y \in R_\Delta(T,x) = R_\Delta(T,x(T))$. Then, for any $t \in [T - \Omega^\flat(T,x,y),T]$, y belongs to $R_\Delta(t,x(t))$. In other words, for every $t \in [T - \Omega^\flat(T,x,y),T]$, $(t,x(t),y) \in Graph(R_\Delta)$ is viable, or, equivalently, the graph of R_Δ is viable under the characteristic system (2.2), p. 32.*

Proof. This is obvious, since the evolution $x(t) = x - y(T-t)$ starting at $t^\sharp := T - \Omega^\flat$ reaches $x(t)$ at time t and starts from $(t^\sharp, \mathbf{k})$, so that $y \in R_\Delta(t,x(t))$. ■

This viability property implies formally that, whenever $(t, x) \rightsquigarrow R_{\Delta}(t, x) = \{r(t, x)\}$ is single-valued and differentiable, it is the solution to the *Burgers partial differential equation*

$$\frac{\partial r(t,x)}{\partial t} + r(t,x)\frac{\partial r(t,x)}{\partial x} = 0 \tag{2.12}$$

satisfying the Dirichlet condition

$$\forall\, t \in [t^{\sharp}, T], \;\; r(t, \mathbf{k}) = \Delta(t) \tag{2.13}$$

so that the endowment threshold function $t \mapsto \Delta(t)$ plays the role of a *Dirichlet boundary condition* in the language of partial differential equations.

Indeed, differentiating the viability relation $r(t, x(t)) = y$ with respect to t, we obtain,

$$\forall\, t \in [t^{\sharp}, T], \;\; \frac{\partial r(t,x(t))}{\partial t} + r(t,x(t))\frac{\partial r(t,x(t))}{\partial x} = 0$$

The Burgers partial differential equation with Dirichlet condition is the cornerstone of the *theory of shock waves*, at the origin of a myriad investigations. We refer only at [64, *Partial Differential Equations*] by *Craig Evans* for a mathematical exposition and to [96, *History of Shock Waves, Explosions and Impact*] by *Peter Krehl* for a physicist one.

Set-valued analysis defining "graphical derivatives" of set-valued maps can be used to give a rigorous mathematical meaning to the concept of a "derivative of a set-valued map," and consequently, to the concept of "set-valued solution" to the Burgers equation. We thus derive that the set-valued map R_{Δ} is the *unique solution with closed graph* satisfying the Burgers equation with Dirichlet condition.

These partial differential equations are just mentioned for those familiar with this theory. Chapter 16 of [15, Aubin, Bayen & Saint-Pierre] provides details on the viability approach for solving conservation laws that it is not useful to reproduce here.

2.5 Providing Commodity

In this case, we assume that the endowment provider furnishes not only an endowment $\Delta(t^{\sharp})$ when the commodity reaches its threshold level $\mathbf{k}$, but also an amount $\Xi(t^{\sharp}) > \mathbf{k}$ through a *threshold commodity function* Ξ.

If the amount $\Xi(t^{\sharp}) > x$ is larger than the objective x at time T, then the agent must refund the provider with some (negative) endowment. If we accept this situation, we have to assume that the endowment function takes also negative values.

Proposition 2.5.1. *[The Threshold Endowment and Commodity Regulation Map] The viability threshold endowment function Δ and the commodity function Ξ being given, the associated regulation map R_Δ providing the endowment and the commodity necessary to reach x at time T in a viable way is the subset $R_{\Delta,\Xi}(T,x)$ of endowments $y = \Delta(T-\Omega^\flat)$ such that the aperture $\Omega^\flat$ is a solution to the equation*

$$\Xi(T-\Omega^\flat) + \Omega^\flat \Delta(T-\Omega^\flat) \;=\; x \tag{2.14}$$

so that the endowments y are given by

$$y \;\in\; \Delta(\Xi^{-1}(x-\Omega^\flat y)) \tag{2.15}$$

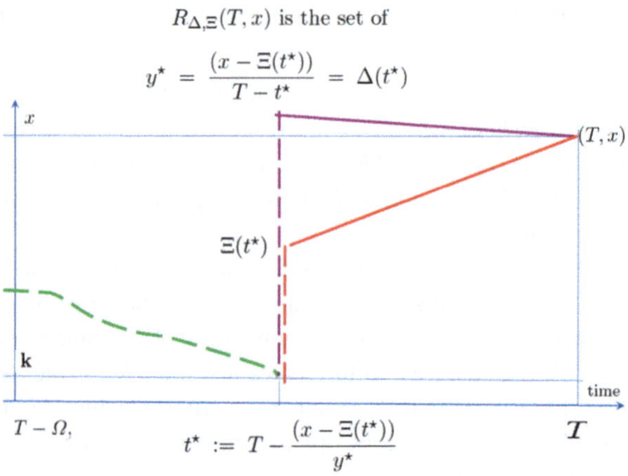

Proof. The proof if a simple adaptation of Proposition 2.3.1, p. 37, where we replace **k** by $\Xi(t^\star)$ in the formula $\Delta\left(T - \frac{(x-\Xi(t^\star))}{y^\star}\right)$ providing the endowment for this new threshold where $t^\star = T - \frac{(x-\Xi(t^\star))}{y^\star}$. ∎

2.6 Towards Viability Solutions

We reformulate the proof in a reverse way for showing that $(T,x) \rightsquigarrow R_\Delta(T,x)$ is the viability solution in a sense that we describe now.

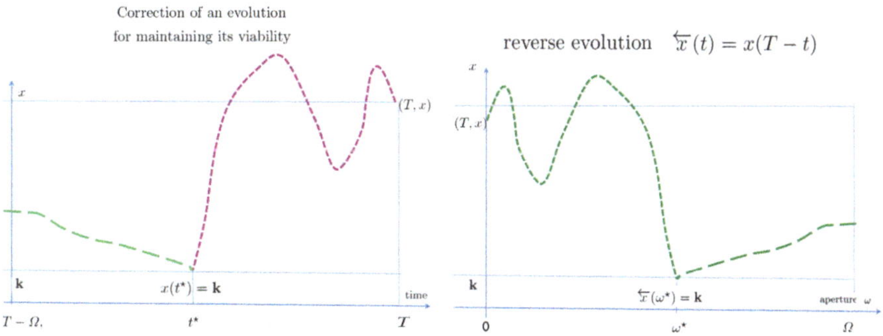

16 ***[Correction of a Evolution To Remain Viable]*** *Left, the temporal graph of an evolution $x(\cdot)$ corrected when it bumps the threshold. Right, its reverse $\overline{x}(\cdot)$.*

Let us introduce the reverse functions $\tau(t) := T - t$, $\overleftarrow{x}(t) := x(T-t) = x - yt$ and $\overleftarrow{y}(t) := y(T-t) = y$ (see Fig. 16, p. 43). We observe that this evolution is governed by the reverse characteristic system

$$\begin{cases} (i) & \tau'(t) = -1 \\ (ii) & \overleftarrow{x}'(t) = -\overleftarrow{y}(t) \\ (iii) & \overleftarrow{y}'(t) = 0 \end{cases} \tag{2.16}$$

starts at (T, x, y) at initial time 0.

It satisfies the constraint $T - t \geq 0$ and $x - yt \geq \mathbf{k}$, i.e., that it is viable in the environment $\mathscr{K} := \mathbb{R}_+ \times [\mathbf{k}, +\infty[\times\mathbb{R}_+$. To say that $y \in \Delta(T - \Omega^\flat)$ whenever $x(T - \Omega^\flat) = \mathbf{k}$ amounts to saying that $(T - \Omega^\flat, x - y(T - \Omega^\flat), y) = (0, \mathbf{k}, y)$ and $y = \Delta(T - \Omega^\flat)$.

Therefore, we introduce

1. the *environment* $\mathscr{K} := \mathbb{R}_+ \times [\mathbf{k}, +\infty[\times\mathbb{R}_+$;
2. the *target* $\mathscr{C} := \{(t, \mathbf{k}, \Delta(t))_{t\geq 0}\}$,

to say that $y \in R_\Delta(T, x)$ amounts to saying that (T, x, y) *is the initial condition of the solution* $t \mapsto (T - t, x - yt, y)$ *governed by the reverse characteristic system (2.16), p. 43,* viable *in the environment* $\mathscr{K}$ *until it* captures *the target* $\mathscr{C}$ *at time* $\Omega^\flat \in [0, \Omega]$.

Why this little awkward hard work while we already solved explicitly the problem in Proposition 2.5.1, p. 42?

The answer is provided in terms of *viable capture basin*[38–124], since we have proved that y belongs to $R_\Delta(T, x)$ if and only if (T, x, y) belongs to the capture basin $\text{Capt}_{(2.16)}(\mathscr{K}, \mathscr{C})$. This is in this sense that R_Δ is a (graphical) viability solution:

$$\text{Graph}(R_\Delta) = \text{Capt}_{(2.16)}(\mathscr{K}, \mathscr{C}) \tag{2.17}$$

2.7 The Viabilist Strategy

The point is that, except in this simplest example, the extension of Proposition 2.5.1, p. 42 to more general problems is difficult: even in this simplest case, the computation of the solution amounts to find the set of fixed points of the nonlinear equation $y = \Delta\left(T - \frac{(x - \mathbf{k})}{y}\right)$, which transfers the difficulties to the (big) ones concerning *the resolution of nonlinear equations or inclusions*.

However, by now, many properties of viable capture basin are known (many of them are gathered in [15, Aubin, Bayen & Saint-Pierre]). Among them are viability kernel algorithms, examples of algorithms computing the viable capture basins. Viability kernel algorithms are part of "set-valued numerical analysis" handling the computations of subsets instead of plain vectors at each step of the algorithm. They provide the numerical solutions, whatever is the function ρ in the characteristic system (2.2), p. 32, the time-dependent viability threshold and the threshold endowment condition. This is how we computed the solution of the example (Fig. 15, p. 40), instead of using algorithms for solving partial differential equations.

This set-valued approach is used pervasively: maps, single-valued or set-valued, are characterized by their graphs and numerical functions by their epigraphs. Closedness of viable capture basins, graphs, and epigraphs is the minimal "continuity property" we require.

Now that this example has been carried over, we shall solve all other problems by defining viability solutions for adequate characteristic systems, constraints, and targets.

Whenever possible, we provide (implicit) analytical formulas characterizing the viability solution and derive the regulation maps. Not because they help us to compute the solutions, but since they tell tales about the portfolio, unfortunately in an esoteric mathematical language.

These viability solutions are also solutions of partial differential equations or inclusions. We shall not insist on this side of the story for lack of (apparent) economic relevance (just the opposite approach of the physicists who use these partial differential inclusions as the starting point of their study and working much harder than we will do for studying their solutions).

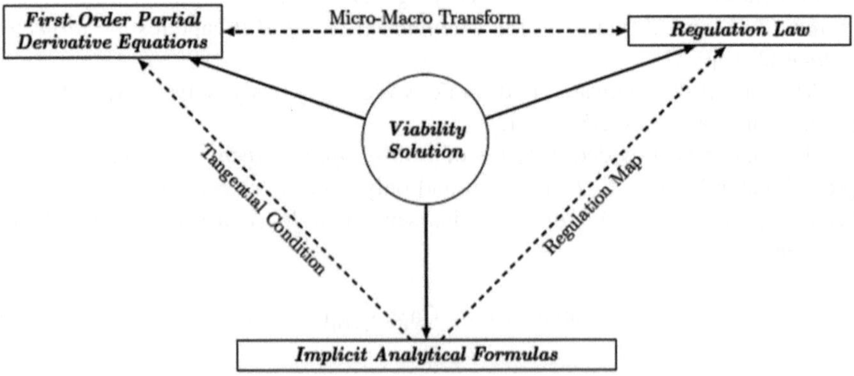

17 ***[Viability Solution: The Mother of Solutions to Three Problems]*** *This diagram describes the three problems under investigation: the underlying invariance property, the microscopic version dealing with the regulation laws and the macroscopic approach through first-order partial differential equation (conservation laws and Hamilton–Jacobi equations). The tools of viability theory allow us to show that the viability solutions solve these three other problems at once and, often, under weaker assumptions.*

Chapter 3
Keeping the Endowment Above a Viability Threshold

From where the provider find the endowment for financing the evolution of the commodity of our solitary economic agent? In Chap. 2, p. 31, we eluded this question by assuming that, at each scenario, it offered a constant endowment y to the agent. This static behavior was described by the choice of the differential equation $y'(t) = 0$.

In this chapter, we neglect agent behavior and concentrate on the really dynamical behavior of the endowment provider described by an affine differential equation

$$y'(t) \;=\; \mathbf{m}(t)y(t) + \mathbf{l}(t)$$

involving

1. an extended function $\mathbf{l}(\cdot) : \mathbb{R} \mapsto \mathbb{R} \cup \{+\infty\}$ representing the *transaction cost function* (simple interests);
2. a *rate of return* $\mathbf{m}(\cdot)$ defining a *discount factor* $M(t) := \int_0^t \mathbf{m}(\tau)d\tau$ (compounded interests for taking into account the curse of the exponential, given here as an assumption, not as the solution to an auxiliary problem describing the necessity of using rates of returns).

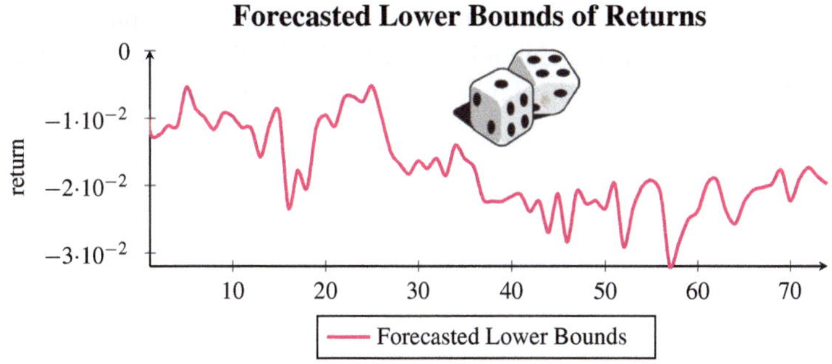

J.-P. Aubin, *Time and Money*, Lecture Notes in Economics and Mathematical Systems 670, DOI 10.1007/978-3-319-00005-3_3,

18 ***[Example of a Provider Behavior]*** *The provider is assumed to be the Stock Market. We extrapolated (with the VIMADES extrapolator*[33–83]*) the (lower bound) of rates of return* $\mathbf{m}(t)$ *of the CAC-40 evolution during the 2011 crisis. It is involved in the basic differential* $y'(t) = (\mathbf{m}_0 + a(t)(\mathbf{m}(t) - \mathbf{m}_0(t))y(t))$ *met in finance governing the value or a portfolio (see [17,* Tychastic Measure of Viability Risk. A Viabilist Portfolio Performance and Insurance Approach*]).*

We also impose time-dependent minimal endowment threshold $\mathbf{k}(t)$ above which the endowment must remain.

Before entering the mathematics, we provide the following illustration:

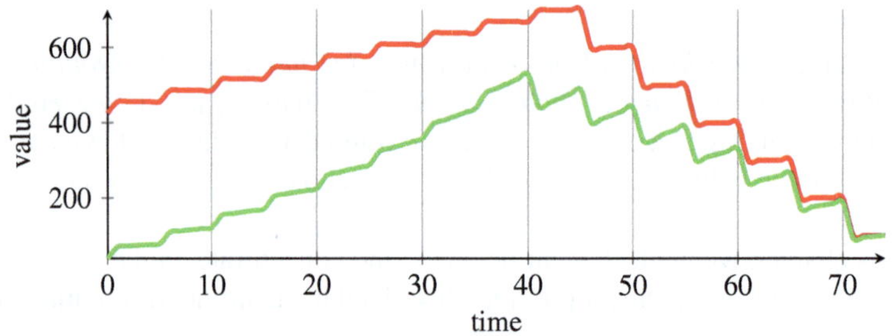

19 ***[Example of Endowment]*** *The graph of the function below denotes a given minimal endowment threshold function* $\mathbf{k}$ *describing a "variable annuities insurance": periodic payments are made during an accumulation phase and periodic payments are obtained during the payout phase. It is no longer continuous, but "punctuated by impulses" at the dates of the schedule. The problem is to compute the* endowment function *above the minimal endowment threshold governed by a differential equation and ending at the value imposed by the threshold. It is computed with a viability algorithm and is displayed above the minimal threshold as it is explained in this chapter.*
Its value at initial time can be regarded as a solvency capital requirement *(SCR) as required in "Pillar I" of the Solvency II framework of the European Directive 2009/138/EC:* "The SCR should reflect a level of eligible own funds that enables insurance and reinsurance undertakings to absorb significant losses and that gives reasonable assurance to policyholders and beneficiaries that payments will be made as they fall due."

3.1 Deterministic Endowment Function

We consider

1. a *threshold function* $\mathbf{k}(\cdot) : \mathbb{R}_+ \mapsto \mathbb{R} \cup \{+\infty\}$;
2. a prescribed aperture $\Omega > 0$ of the temporal window $[T - \Omega, T]$.

Definition 3.1.1. [The Endowment Function] The *endowment function* $W : T \mapsto W(T) \in \mathbb{R}$ is defined at time T by

$$W(T) := \sup_{t \in [T-\Omega, T]} \left(e^{\int_t^T \mathbf{m}(\sigma)d\sigma} \mathbf{k}(T-t) + \int_t^T e^{\int_\tau^T \mathbf{m}(\sigma)d\sigma} \mathbf{l}(\tau)d\tau \right) \tag{3.1}$$

3.2 Viability Solution

For proving that the endowment function is the viability solution that we define next, we introduce reverse evolutions as it was explained in Sect. 2.6, p. 42, and Fig. 16, p. 43.

We associate with any function $\mathbf{v} : t \in \mathbb{R}_+ \mapsto \mathbf{v}(t) \in \mathbb{R} \cup \{+\infty\}$ the function $\overleftarrow{\mathbf{v}}(t) \in \mathbb{R}_+ \mapsto \mathbb{R} \cup \{+\infty\}$ defined by

$$\forall\, t \leq T, \;\; \overleftarrow{\mathbf{v}}(t) \;:=\; \mathbf{v}(T-t)$$

where T is not indicated for the rigorous definition $\overleftarrow{\mathbf{v}}_T(\cdot)$ for simplifying formulas. Conversely, we can recover $\mathbf{v}(\cdot)$ from $\overleftarrow{\mathbf{v}}(\cdot)$ since

$$\forall\, t \geq 0, \;\; \mathbf{v}(t) \;:=\; \overleftarrow{\mathbf{v}}(T-t)$$

Observe also that

$$\int_0^t \overleftarrow{\mathbf{v}}(\tau)d\tau \;=\; \int_{T-t}^T \mathbf{v}(\sigma)d\sigma \tag{3.2}$$

We associate with the *rate of return* $\mathbf{m}(\cdot)$ its reverse rate of return $\overleftarrow{\mathbf{m}}(\cdot)$ and its *reverse discount factor* $\overleftarrow{M}(\cdot)$ defined by

$$\overleftarrow{M}(t) \;:=\; \int_0^t \overleftarrow{\mathbf{m}}(\tau)d\tau \;=\; \int_{T-t}^T \mathbf{m}(\sigma)d\sigma \;=\; M(T) - M(T-t) \tag{3.3}$$

where $M(t) := \displaystyle\int_0^t \mathbf{m}(\sigma)d\sigma$ is the discount factor.

The function $y(\cdot)$ defined by

$$y(t) = e^{M(t)}\left(e^{-M(T)}y - \int_t^T e^{-M(\tau)}\mathbf{l}(\tau)d\tau\right) \tag{3.4}$$

is the solution to the differential equation $y'(t) = \mathbf{m}(t)y(t) + \mathbf{l}(t)$ *arriving* at y at time T if and only if $\overleftarrow{y}(\cdot)$ defined by

$$\overleftarrow{y}(t) = e^{-\overleftarrow{M}(t)}\left(y - \int_0^t e^{\overleftarrow{M}(\tau)}\overleftarrow{\mathbf{l}}(\tau)d\tau\right) \tag{3.5}$$

is the solution to the reverse differential equation $\overleftarrow{y}'(t) = -\overleftarrow{\mathbf{m}}(t)\overleftarrow{y}(t) - \overleftarrow{\mathbf{l}}(t)$ *starting* at y at time 0.

Let us consider the reverse *"characteristic system"*

$$\begin{cases} (i) \;\; \overleftarrow{\tau}'(t) = -1 \\ (ii) \;\; \overleftarrow{y}'(t) = -\overleftarrow{\mathbf{m}}(t)\overleftarrow{y}(t) - \overleftarrow{\mathbf{l}}(t) \end{cases} \tag{3.6}$$

the solution $(T - t, \overleftarrow{y}(\cdot))$ of which starting at $(0, y)$ at initial time 0 is equal to

$$\begin{cases} \forall t \geq 0 \\ (i) \;\; \overleftarrow{\tau}(t) = T - t \\ (ii) \;\; \overleftarrow{y}(t) = e^{-\overleftarrow{M}(t)}\left(y - \int_0^t e^{\overleftarrow{M}(\tau)}\overleftarrow{\mathbf{l}}(\tau)d\tau\right) \end{cases} \tag{3.7}$$

We associate with $\mathbf{k}(\cdot)$ and Ω the function $\mathbf{c} : t \rightsquigarrow \mathbf{c}(t) \in \mathbb{R} \cup \{+\infty\}$ defined by

$$\mathbf{c}(t) := \begin{cases} \mathbf{k}(T - \Omega) & \text{if } t = T - \Omega \\ +\infty & \text{if } t \neq T - \Omega \end{cases} \tag{3.8}$$

Definition 3.2.1. [**The Viability Solution**] The viability solution $V(T) \in \mathbb{R} \cup \{+\infty\}$ at time T is defined by

$$V(T) := \inf_{(T,y)\in \mathrm{Capt}_{(3.6)}(\mathscr{E}p(\mathbf{k}),\mathscr{E}p(\mathbf{c}))} y \tag{3.9}$$

As announced, the endowment function is equal to the valuation solution, which enjoys all properties of viable capture basins, some of them presented here. Since "viability algorithms" have been designed to compute viable capture basins, they can be used to calculate endowment functions as shown in Fig. 19, p. 48.

Theorem 3.2.2. *[The Viability Solution is the Endowment Function] The* viability volution V *defined by (3.9), p. 50, is equal to the* endowment function $W(\cdot)$ *defined by (3.1), p. 49:*

$$\mathscr{E}p(W) \;=\; \mathscr{E}p(V) \;=\; \mathrm{Capt}_{(3.6)}(\mathscr{E}p(\mathbf{k}), \mathscr{E}p(\mathbf{c})) \tag{3.10}$$

In other words, the epigraph of the endowment function is the capture basin of the epigraph of the objective cost function viable in the epigraph of the threshold function.

Proof.

1. For proving the theorem, we perform the change of the variable $s = T - t$ and recall that $M(T) - M(t) = \overleftarrow{M}(T-t)$ and that $D =: T - \Omega$. Hence the formula defining $W(T)$ in the form (3.1), p. 49, can be rewritten as

$$W(T) \;=\; \sup_{t\in[0,t^\star]} \left(e^{\overleftarrow{M}(t)}\mathbf{k}(T-t) + \int_0^t e^{\overleftarrow{M}(\tau)}\overleftarrow{\mathbf{I}}(\tau)d\tau \right)$$

2. Let (T, y) belong to the capture basin $\mathrm{Capt}_{(3.6)}(\mathscr{E}p(\mathbf{k}), \mathscr{E}p(\mathbf{c}))$ and consider the solution made of $t \mapsto \overleftarrow{\tau}(t) = T - t$ and

$$t \mapsto \overleftarrow{y}(t) \;=\; e^{-\overleftarrow{M}(t)} \left(y - \int_0^t e^{\overleftarrow{M}(\tau)}\overleftarrow{\mathbf{I}}(\tau)d\tau \right)$$

to the characteristic system (3.6) starting at (T, y). By definition of the capture basin, there exists a finite time $t^\star \geq 0$ such that

a. $(\overleftarrow{\tau}(t^\star), \overleftarrow{y}(t^\star))$ belongs to the epigraph of $\mathbf{c}$, which means that

$$\mathbf{c}(T-t^\star) \leq \overleftarrow{y}(t^\star) \;=\; e^{-\overleftarrow{M}(t^\star)} \left(y - \int_0^{t^\star} e^{\overleftarrow{M}(\tau)}\overleftarrow{\mathbf{I}}(\tau)d\tau \right) \tag{3.11}$$

i.e., by the very definition (3.8), p. 50, of the function $\mathbf{c}(\cdot)$, that $t^\star = \Omega$ and that

$$e^{\overleftarrow{M}(\Omega)}\mathbf{k}(T-\Omega) + \int_0^{\Omega} e^{\overleftarrow{M}(\tau)}\overleftarrow{\mathbf{I}}(\tau)d\tau \;\leq\; y \tag{3.12}$$

b. The solution is viable in the epigraph of the constraint transaction on the interval $[0, \Omega]$, i.e.,

$$\sup_{t\in[0,\Omega]} \left(e^{\overleftarrow{M}(t)}\mathbf{k}(T-t) + \int_0^t e^{\overleftarrow{M}(\tau)}\overleftarrow{\mathbf{I}}(\tau)d\tau \right) \;\leq\; y \tag{3.13}$$

Hence, putting together these two inequalities, we obtain

$$y \geq W(T) = \sup_{t \in [0,\Omega]} \left(e^{\overleftarrow{M}(t)} \mathbf{k}(T-t) + \int_0^t e^{\overleftarrow{M}(\tau)} \overleftarrow{\mathbf{l}}(\tau) d\tau \right)$$

3. Taking the infimum over the y such that $(T, y) \in \text{Capt}_{(3.6)}(\mathscr{E}p(\mathbf{k}), \mathscr{E}p(\mathbf{c}))$, we infer that the viability solution $W(T) \leq V(T)$.
4. Conversely, let us consider the solution defined by $\overleftarrow{\tau}(t) = T - t$ and

$$\overleftarrow{y}(t) = e^{-\overleftarrow{M}(t)} \left(W(T) - \int_0^t e^{\overleftarrow{M}(\tau)} \overleftarrow{\mathbf{l}}(\tau) d\tau \right) \tag{3.14}$$

to the characteristic system (3.6) starting at $(T, W(T))$ at initial time 0.

By the definition of $W(T)$, we observe that $\overleftarrow{y}(\Omega) \geq \mathbf{c}(T - \Omega)$ and $\overleftarrow{y}(t) \geq \mathbf{k}(T - t)$ for all $t \in [0, \Omega]$. This means that $(T, W(T))$ belongs to the capture basin $\text{Capt}_{(3.6)}(\mathscr{E}p(\mathbf{k}), \mathscr{E}p(\mathbf{c}))$. By definition of the viability solution, $V(T) \leq W(T)$.

Hence equality $V(T) = W(T)$ ensues. ■

Therefore, for every $t \in [\Omega, T]$, one can define the viability solution $V(t)$. We shall derive the explicit formula of $V(t)$ in function of $W(T)$ from the bilateral fixed point property (see Theorem 8.2.1, p. 122).

Theorem 3.2.3. *[Analytic Formula of the Viability Solution] The viability solution $t \mapsto V(t)$ is equal to the solution*

$$\forall\, t \in [T - \Omega, T], \;\; V(t) = e^{M(t)} \left(e^{-M(T)} W(T) - \int_t^T e^{-M(\tau)} \mathbf{l}(\tau) d\tau \right) \tag{3.15}$$

to the differential equation

$$V'(t) = \mathbf{m}(t) V(t) + \mathbf{l}(t) \text{ arriving at } W(T) \text{ at time } T \tag{3.16}$$

In other word, knowing the endowment function at final time T, equation (3.15) provides the endowment function $W(t)$ for earlier times $t \in [T - \Omega, T]$, which can be obtained solving differential equation (3.16) in reverse time from the final condition $V(T) = W(T)$.

Proof. By Theorem 8.2.5, p. 123, we know that the capture basin $\mathscr{E}p(V) = \text{Capt}_{(3.6)}(\mathscr{E}p(\mathbf{k}), \mathscr{E}p(\mathbf{c}))$ of the epigraph of $\mathbf{c}$ under the auxiliary system (3.6), p. 50, is the bilateral fixed point

$$\text{Capt}_{(3.6)}(\mathscr{E}p(V), \mathscr{E}p(\mathbf{c})) = \mathscr{E}p(V) = \text{Capt}_{(3.6)}(\mathscr{E}p(\mathbf{k}), \mathscr{E}p(V))$$

Since $(T, W(T)) \in \mathscr{E}p(V)$ belongs to the capture basin $\mathrm{Capt}_{(3.6)}(\mathscr{E}p(V), \mathscr{E}p(\mathbf{c}))$ of the epigraph of $\mathbf{c}$ viable in $\mathscr{E}p(V)$ under the auxiliary system, inequality (3.13), p. 51, of the proof of Theorem 3.2.2, p. 51 where we replace $\mathbf{k}$ by V implies that there exists $\Omega \geq 0$ such that, in particular,

$$\forall\, t \in [0, \Omega], \quad V(T-t) \leq \overleftarrow{y}(t) = e^{-\overleftarrow{M}(t)}\left(W(T) - \int_0^t e^{\overleftarrow{M}(\tau)}\overleftarrow{\mathbf{l}}(\tau)d\tau\right) \tag{3.17}$$

We shall deduce the opposite inequality

$$\forall\, t \in [0, \Omega], \quad V(T-t) \leq \overleftarrow{y}(t) \tag{3.18}$$

from the second fixed point property stating that $\mathscr{E}p(V) = \mathrm{Capt}_{(3.6)}(\mathscr{E}p(\mathbf{k}), \mathscr{E}p(V))$.

Since the viability solution is defined by $V(T) := \inf_{(T,y)\in \mathrm{Capt}_{(3.6)}(\mathscr{E}p(\mathbf{k}),\mathscr{E}p(V))} y$ and since we assumed that $(T, V(T))$ belongs to $\mathrm{Capt}_{(3.6)}(\mathscr{E}p(\mathbf{k}), \mathscr{E}p(V))$, whenever $V(T) > \overleftarrow{k}(0)$, we deduce that for any $\varepsilon > 0$ such that $\varepsilon < V(T) - \overleftarrow{k}(0)$, $(T, V(T) - \varepsilon)$ does not belong to $\mathrm{Capt}_{(3.6)}(\mathscr{E}p(\mathbf{k}), \mathscr{E}p(V))$. By definition of capture basins, this implies that the evolution $t \mapsto (T - t, \overleftarrow{x}(t), \overleftarrow{z}_\varepsilon(t))$ governed by the characteristic system (3.6), p. 50 starting at $(T, V(T) - \varepsilon)$ remain outside of the target $\mathscr{E}p(V)$ forever or until some $\delta_\varepsilon > 0$ when $\overleftarrow{z}_\varepsilon(\delta_\varepsilon) = \mathbf{k}(T - \delta_\varepsilon, \overleftarrow{x}(\delta_\varepsilon))$.

Let

$$\overleftarrow{y}(t) := e^{-\overleftarrow{M}(t)}\left(W(T) - \int_0^t e^{\overleftarrow{M}(\tau)}\overleftarrow{\mathbf{l}}(\tau)d\tau\right)$$

and

$$\overleftarrow{z}_\varepsilon(t) := \overleftarrow{y}(t) - \varepsilon e^{-\overleftarrow{M}(t)} \tag{3.19}$$

Therefore, $z_\varepsilon(\cdot)$ is governed by the characteristic system (3.6), p. 50 starting at $(T, V(T) - \varepsilon)$, which remains outside the epigraph of V forever or until some $\delta_\varepsilon \in]0, T]$ when $z_\varepsilon(\delta_\varepsilon) = \overleftarrow{y}(\delta_\varepsilon) - \varepsilon e^{-\overleftarrow{M}(\delta_\varepsilon)} = \mathbf{k}(T - \delta_\varepsilon)$.

This implies that

$$\forall\, t \in [0, \delta_\varepsilon], \quad \overleftarrow{z_\varepsilon}(t) < V(T-t) \tag{3.20}$$

Let δ be the limit of some subsequence δ_ε when $\varepsilon \to 0+$, which satisfies $\overleftarrow{y}(\delta) = \mathbf{k}(T - \delta)$. We deduce that

$$\forall\, t \in [0, \delta], \quad \overleftarrow{y}(t) \leq V(T-t) \tag{3.21}$$

and thus, equality

$$\forall\, t \in [0, \delta], \quad \overleftarrow{y}(t) = V(T-t, \overleftarrow{x}(t)) \tag{3.22}$$

Therefore, $V(T-t) = \overleftarrow{y}(t)$, and consequently, $V(t) = y(t)$ is the solution

$$V(t) \;=\; e^{M(t)}\left(e^{-M(T)}W(T) - \int_t^T e^{-M(\tau)}\mathbf{l}(\tau)d\tau\right)$$

to the differential equation $V'(t) = \mathbf{m}(t)V(t) + \mathbf{l}(t)$ arriving at $W(T)$ at time T. ■

3.3 Correction of Unviable Potential Functions

Let us consider a given *potential function* $U(\cdot)$ that does not satisfy necessarily the threshold $U(t) \geq \mathbf{k}(t)$ for all $t \in [T-\Omega, T]$. The question arises to devise a strategy to correct it by a viable endowment function $W(\cdot)$:

Proposition 3.3.1. *[**Correction of Unviable Potential Functions**]Let us consider a (differentiable) potential function $U(\cdot)$ which is not necessarily viable. The corrected endowment function W_U is defined by*

$$\left\{\begin{array}{l} W_U(T) + U(T) \;:= \\ \sup\limits_{t \in [T-\Omega, T]} \left(e^{\int_t^T \mathbf{m}(\sigma)d\sigma}[\mathbf{k}(t) + U(t)] + \int_t^T e^{\int_\tau^T \mathbf{m}(\sigma)d\sigma}\mathbf{l}(\tau)d\tau\right) \end{array}\right. \tag{3.23}$$

It amounts to add the potential function $U(t)$ to both the threshold function $\mathbf{k}$ *and the function* $\mathbf{l}$ *and subtract the final value $U(T)$. The corrected endowment function $t \mapsto W_U(t)$ is equal to the solution: $\forall\, t \in [T-\Omega, T]$,*

$$W_U(t) \;=\; e^{-\int_t^T \mathbf{m}(\sigma)d\sigma}[W(T) + U(T)] - \int_t^T e^{\int_\tau^t \mathbf{m}(\sigma)d\sigma}\mathbf{l}(\tau)d\tau - U(t) \tag{3.24}$$

to the differential equation

$$W_U'(t) \;=\; \mathbf{m}(t)(W_U(t) + U(t)) + \mathbf{l}(t) - U'(t) \tag{3.25}$$

arriving at $W(T) + U(T)$ at time T.

Proof. It is sufficient to introduce the Lagrangian associated with U by the formula

$$\mathbf{l}_U(t) \;:=\; \mathbf{m}(t)U(t) - U'(t) \tag{3.26}$$

which implies that

$$\int_t^T e^{-M(\tau)} \mathbf{l}_U(\tau) d\tau = e^{M(t)} U(t) - e^{M(T)} U(T) \qquad (3.27)$$

thanks to formula (3.4), p. 50.

Therefore, adding to the initial transaction cost function $\mathbf{l}$ the Lagrangian $\mathbf{l}_U$ associated with the *potential function* $U(\cdot)$, formula (3.1), p. 49, becomes formula (3.23), p. 54. The other formulas follow from Theorem 3.2.3, p. 52. ■

Chapter 4
Uncertain Endowments and Economic Cycles

This chapter concerns the case when the functions $\mathbf{l}(t)$ (simple interests) or the rate of returns $\mathbf{m}(t)$ (compound interests) *are no longer given*, as in Chap. 3, p. 47, but are themselves solutions to a viability problem requiring that the endowment evolves between a lower *viability threshold* $\mathbf{k}$ (for the agent) and upper *scarcity threshold* $\mathbf{c}$ (for the provider), assumed to be constant for the simplicity of the exposition:

$$\forall\, t \geq 0,\ \ \mathbf{k}\ \leq\ y(t)\ \leq\ \mathbf{c} \tag{4.1}$$

Always for the sake of simplifying the exposition, we shall split the differential equation

$$y'(t)\ =\ \mathbf{m}(t)y(t) + \mathbf{l}(t)$$

into two differential equations:

1. **Simple Interests.**

$$\forall\, t \geq 0,\ \ y'(t)\ =\ \mathbf{l}(t) \text{ where } \mathbf{l}(t)\ \in\ \mathbb{R} \tag{4.2}$$

2. **Compounded Interests.**

$$\forall\, t \geq 0,\ \ y'(t) = \mathbf{m}(t)y(t) \text{ where } \mathbf{m}(t)\ \in\ \mathbb{R} \tag{4.3}$$

where the interest rates $\mathbf{l}(t)$ and $\mathbf{m}(t)$ are chosen at each time t for governing evolutions confined in the interval $[\mathbf{k}, \mathbf{c}]$. This will allows us t*o compare the consequences of the choice between simple and compounded interests.*

We denote by r either one of the simple rate $r = \mathbf{l}$ or compounded rate $r = \mathbf{m}$ any time when the comments are valid for both cases: we shall observe that the results share the same *qualitative* properties, but not the *quantitative* ones.

J.-P. Aubin, *Time and Money*, Lecture Notes in Economics and Mathematical Systems 670, DOI 10.1007/978-3-319-00005-3_4,

Using compounded interests instead of simple ones in the dynamical analysis of the endowment provides exponential evolutions, which may not always be in tune with economic growth. This is however such an ancient tradition going back to the discovery of compounded interests, that there is no realistic hope to reverse to simple interests for regulating the evolution of endowment values (see Sect. 1.3, p. 13): this would be a situation where returning so long ago is not reactionary, but (r)evolutionary.

20 ***[Assumption: Bounding Interests by an Inertia Threshold]*** *We assume that simple or compounded interest rates are not known, but are only bounded by an* inertia threshold $0 \leq c < +\infty$*:*

$$\forall\, t \in \mathbb{R}^{+}, \quad |\mathbf{l}(t)| \leq c \text{ (respectively } |\mathbf{m}(t)| \leq c) \tag{4.4}$$

Therefore, the endowment laws are no longer deterministic, but open to (contingent) uncertainty to locate what are viable interest rates to be chosen for keeping the endowments viable. The problem is to find the *regulation map* $y \rightsquigarrow \mathbf{l} \in R_c(y)$ and $y \rightsquigarrow \mathbf{m} \in R_c(y)$ respectively for governing viable evolutions of the endowment.

4.1 Inertia Principle and Heavy Evolutions

Next, once this task is completed, *the question arises: how can we find those interest rates for the evolution of the endowment to be viable?*

We have indeed to find single-valued selections (feedbacks) $y \mapsto r_c(y) \in R_c(y)$ providing the viable simple and compounded interest rates $\mathbf{l} = r_c(y)$ and $\mathbf{m} = r_c(y)$ respectively telling the endowment provider the interest rates to be chosen when the endowment is known.

Among many possible choices, we propose to accept the inertia principle:

21 ***[The Inertia Principle]*** *In a loose way, the* inertia principle *states that the* simple and compounded "interest rates" of the system are kept constant as long as possible and changed only when viability or inertia is at stake.

The inertia principle provides a mathematical explanation of the emergence of the concept of *punctuated equilibrium* introduced in paleontology by *Nils Eldredge* and *Stephen J. Gould* in 1972.

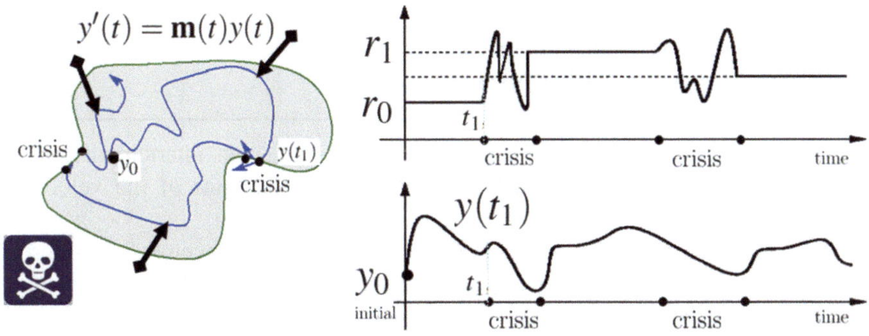

22 ***[Punctuated Evolution]*** *Starting from* y_0 *with the constant interest rate* r_0*, the endowment evolves in* K *until time* t_1*, (first* punctuated equilibrium phase*) when the state* $y(t_1)$ *is about to leave the environment* K *and when the constant interest rate* r_0 must *start to evolve. Then a* critical phase *happens during which velocities also evolve (as slowly as possible) to maintain viability, until time* t_1 *when the control* r_1 *can remain constant during a nonempty time interval:* second punctuated equilibrium phase, *after which starts a second critical phase, etc. In this simple example,* $K = [\mathbf{k}, \mathbf{c}]$ *and* $r \in \mathbb{R}$.

Hence the question arises: *how can we implement the inertia principle*? *When, where, and how* interest rates must cease to be constant and start to "evolve" in order to guarantee both the viability thresholds (4.1), p. 57, of the endowment and the inertia threshold (4.4), p. 58?

There are numerous methods for regulating evolutions satisfying the inertia principle. At this point, we posit a new assumption: *interest rates evolve as "slowly" as possible because their change may be (very) "costly."*

We thus introduce the simplest way to achieve this objective, by selecting at each instant the simple or compounded interest rates providing viable evolutions with *minimal velocity*.

The best candidates are constant interest rates: the endowments are respectively equal to $y(t) = y + \mathbf{l}t$ and $y(t) = ye^{\mathbf{m}t}$: in this case, they reach the upper scarcity threshold $\mathbf{c}$ (when $\mathbf{l} > 0$, $\mathbf{m} > 0$ and $y < \mathbf{c}$) at *"exit times"* $T = \dfrac{\mathbf{c} - y}{\mathbf{l}}$ and $T = \dfrac{\log(\mathbf{c}/y)}{\mathbf{m}}$ respectively after which they cease to be viable.

Therefore, the inertia principle requires that these simple and compounded interest rates have to be changed before reaching these exit times.

Definition 4.1.1. [**Heavy Evolutions of Endowments**] *Heavy evolutions* of endowments are those governed by interest rates $\mathbf{l}(t) \in R_c(y(t))$ and $\mathbf{m}(t) \in R_c(y(t))$ respectively *which minimize at each instant* (the norms of) their velocities. Whenever their velocities are equal to 0 on a temporal window, then they are selected and govern constant interest rates as long as they evolve in such a temporal window. Evolutions obtained in this way are called *"heavy" viable evolutions* in the sense of *heavy trends* in economics.

Minimizing *at each instant* the velocity is no longer an intertemporal minimization problem of a functional cumulating costs functions of the velocities (see Sect. 1.5, p. 20).

We summarize the results of Chaps. 6 and 7 of [15, Aubin, Bayen & Saint-Pierre] for underlying the main points. Before entering the mathematical details, we display graphs of the heavy evolutions:

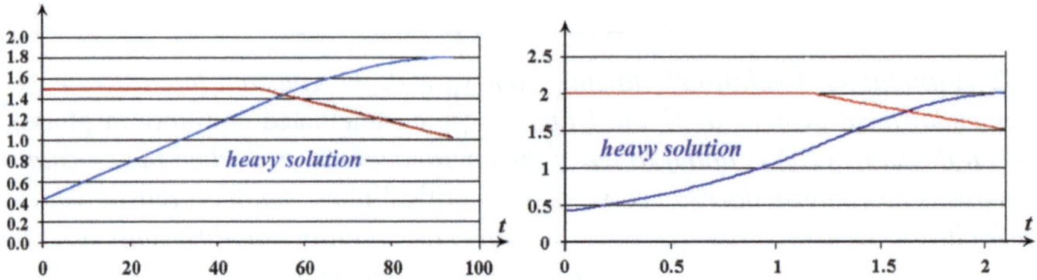

23 ***[Graphs of the Heavy Evolutions]*** *Both the graphs of the heavy evolution and of its interest rate are plotted. They are computed from the Viability Kernel Algorithm. The* interest rate remains constant *until the trajectory of the endowment hits the boundary of the graph of the regulation map. Then the heavy evolution and decreases to the boundary with velocity equal to* $-c$.
[Left] Case of the simple interests: *The evolution* is affine in time *until it hits the boundary and decreases to the boundary.*
[Right] *Case of the compound interests: the heavy evolution* growths exponentially in time *until it hits the boundary and decreases to the boundary.*

For achieving this task, we need to introduce *inertia functions* defined by

1. **Simple Interests**: when the interest rate $\mathbf{I}(t) := y'(t)$ is the velocity, the inertia function is defined by

$$\alpha(y,\mathbf{I}) := \inf_{(y(\cdot),\mathbf{I}(\cdot))\in\mathscr{P}(y,\mathbf{I})} \sup_{t\geq 0} \|\mathbf{I}'(t)\| = \inf_{(y(\cdot),\mathbf{I}(\cdot))\in\mathscr{P}(y,\mathbf{I})} \sup_{t\geq 0} \|y''(t)\| \tag{4.5}$$

where $\mathscr{P}(y,\mathbf{I})$ denote the set of value-interest rate solutions $(y(\cdot),\mathbf{I}(\cdot))$ to differential equation (4.2) viable in the interval $[\mathbf{k},\mathbf{c}]$ such that $y(0) = y$ and $\mathbf{I}(0) = y'(0) = \mathbf{I}$. It is called in this case the *Newtonian inertia function* because, in mechanics, the evolution of the second derivative $y''(t)$, i.e., the acceleration, is inversely proportional to the mass: the smaller the evolution of acceleration, the larger the evolution inertia.

2. **Compounded Interests** We extend this concept of inertial to the velocity of the growth rates for the economic version, where growth rates $\mathbf{m}(t) := \frac{y'(t)}{y(t)}$ replace velocities. The inertia function, called in this case the *exponential inertia function*, is defined by

$$\alpha(y,\mathbf{m}) := \begin{cases} \inf_{(y(\cdot),\mathbf{m}(\cdot))\in\mathscr{P}(y,\mathbf{m})} \sup_{t\geq 0} |\mathbf{m}'(t)| \\ = \inf_{(y(\cdot),\mathbf{m}(\cdot))\in\mathscr{P}(y,\mathbf{m})} \sup_{t\geq 0} \left(\mathbf{m}(t) \left\| \frac{y''(t)}{y(t)} - \mathbf{m}(t) \right\| \right) \end{cases} \tag{4.6}$$

where $\mathscr{P}(y,\mathbf{m})$ the set of solutions $(y(\cdot),\mathbf{m}(\cdot))$ to system (4.3) *viable* in the interval $[\mathbf{k},\mathbf{c}]$ starting at $(y,\mathbf{m})$.

We can characterize these inertia functions by viability kernels and capture basins (see Chaps. 6 and 7 of [15, Aubin, Bayen & Saint-Pierre] for details. The numerical results using viability algorithms are due to Patrick Saint-Pierre). For the time, we provide only the graphs of the inertia functions and their sections:

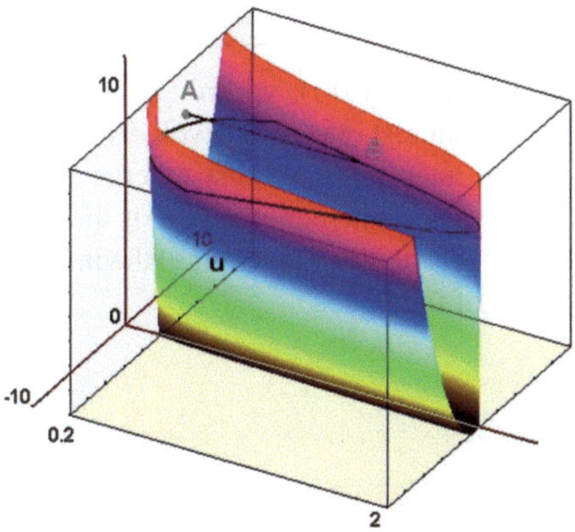

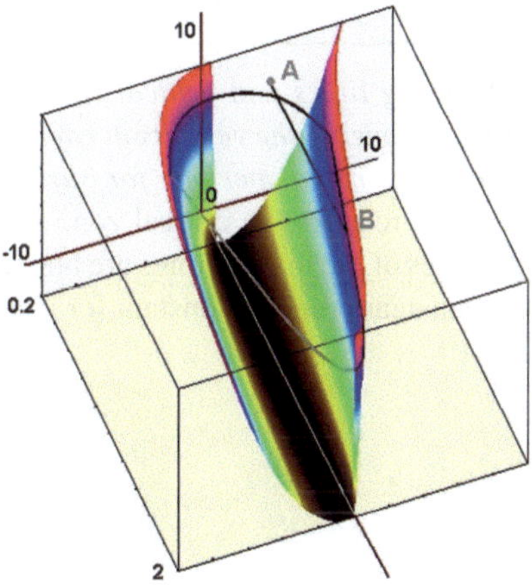

24 ***[Graphs of the Inertia Functions]*** *The Viability Kernel Algorithm computes the graph of the inertia functions.*
[Left] Case of simple interest rate *(Newtonian inertia function).*
[Right] Case of compound interest rate *(exponential inertia function).*

From now on, we shall denote by $r(t)$ either $\mathbf{l}(t)$ or $\mathbf{m}(t)$ since these simple and compounded rates will be used in the same way for studying heavy and cyclic evolutions.

For defining heavy solutions, *we still fix a bound c on the norms of the velocities of the interest rates* and take any initial pair (y,r) such that $\alpha(y,r) < c$. We then fix the interest rate r and consider the evolution $(y_r(t), r)$ under constant interest rate where $y_r(\cdot)$ is the solution to differential equations $y'(t) = r$ or $y'(t) = ry(t)$ (evolving with velocity $r'(t) = 0$).

As long as $\alpha(y_r(t), r)$ is strictly smaller than the velocity bound c, the interest rate r inherited from the past can be maintained, allowing the system to be regulated by this constant interest rate r. Since the value $y_r(\cdot)$ of the system evolves while the interest rate remains constant and equal to r, the inertia function $\alpha(y_r(t), r)$ evaluated on such an evolution may increase and eventually overrun the bound c measuring the maximal velocity of the interest rates at some time $\sigma_c(y, r)$ at the value $y_r(\sigma_c(y, r)) \in \Xi_c(r)$ providing *warning time* defined as follows:

Definition 4.1.2. [**Warning Times**] Knowing the inertia function, assume that $c > \alpha(y, r)$. Then the *warning time* $\sigma_c(y, r) \in \mathbb{R} \cup \{+\infty\}$ is the first instant when the evolution $y_r(\cdot)$ starting from y reaches the *critical zone*

$$\Xi_c(r) := \{y \in \mathrm{Dom}(U) \text{ such that } \alpha(y, r) = c\}$$

of the interest rate r at $y_r(\sigma_c(y, r)) \in \Xi_c(r)$.

Warning times and *critical zones* tell us **when** $(\sigma_c(y, r))$, **where** $(\Xi_c(r))$, and **how** (*minimizing the velocity at each instant*) the interest rates must evolve, defining a *viability critical period: for surviving*, other interest rates must emerge when the value reaches the critical zone of the interest rate, in such a way that the new velocities of the interest rates are bounded by the *inertia threshold c* until the interest rate can again remain constant for a new period of time.

25 *[**Warning Time or "Kairos"**] The concept of* warning time *is a mathematical translation of the anglo-saxon concept of* timing, *or the Italian concept of* trempais, *modernizing the concept of* kairos *of classical Greece, meaning* opportune *moment. The ancient Greeks used this* qualitative *concept of time by opposition to* chronos, *the* quantitative... *chronological time, which can be measured by clocks.*

Lysippos sculptured a wonderful *concrete* representation of this very *abstract* concept, which was depicted by *Politicos* who gave his definition of the kairos of *Lysippos*'s bas-relief[1] (in the museum of Torino).

4.2 Viability Oscillators and Economic Cycles

Economic evolutions offer to our investigations myriads of *economical clocks* or oscillators, producing *periodic evolutions*, or, rather, *cyclic evolutions*. This modification of the terminology is justified by the fact that nowadays, periodic evolutions are understood *as produced by a system of periodic differential equations*. The search of these equations is a very difficult undertaking, and may not be realistic at all. Hence the question arises to look for *other ways to produce cyclic solutions* to underlie the fact that *they are not solutions of a given system of periodic differential equations*.

Instead of guessing complex systems of differential equations for "economical clocks,"[12–23] the underlying assumptions of our engine producing cyclic evolutions is

26 ***[Viability Oscillator]*** *A* viability oscillators *is made of*

1. **viability constraints** *on the value of the system in a given environment,*
2. **inertia thresholds** *imposing a* speed limit *on each component of the evolutions of the interest rates.*

The account of cyclic evolutions we present may illustrate mathematically the emergence of economic cycles highlighted by *Clément Juglar* in the nineteenth century, and, after him, *Nikolai Kondratiev* and others.

[1] *"Who and whence was the sculptor? From Sikyon. And his name? Lysippos. And who are you? Time who subdues all things. Why do you stand on tip-toe? I am ever running. And why you have a pair of wings on your feet? I fly with the wind. And why do you hold a razor in your right hand? As a sign to men that I am sharper than any sharp edge. And why does your hair hang over your face? For him who meets me to take me by the forelock. And why, in Heaven's name, is the back of your head bald? Because none whom I have once raced by on my winged feet will now, though he wishes it sore, take hold of me from behind. Why did the artist fashion you? For your sake, stranger, and he set me up in the porch as a lesson."*

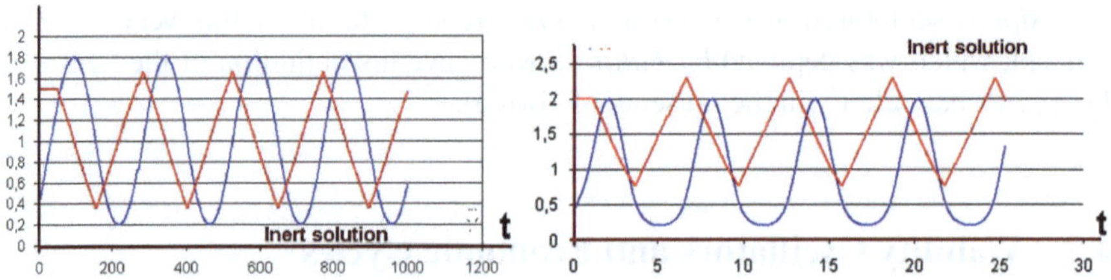

27 Trajectory of Cyclic Evolutions. The evolution oscillates between **k** and **c** back and forth by ranging alternatively two different trajectories on the metaenvironment $[\mathbf{k}, \mathbf{c}] \times \mathbb{R}$. We use only two velocities of the interest rates, $-c$ and $+c$, for regulating the metasystem (instead of an infinite family of open loop interest rates as in the control theory). [Left] Case of the inertia function. [Right] Case of the exponential inertia function.

The cyclic evolutions ranging over the boundary of the regulation map satisfy the *Titanic effect* described by the following property: *a value-interest rate pair* $(y(\cdot), r(\cdot)) \in \mathbb{R}^2$ *satisfies the* Titanic effect *on the interval* I *if*

$$\forall\, t \in I, \quad y'(t)r'(t) \;<\; 0$$

The Titanic effect means that a decrease of the interest rate corresponds to an increase of the value and vice-versa.

An equilibrium state is not necessarily a favorable situation. Why should we remain stuck at equilibrium $(a, 0)$ whereas we may "prefer" a larger value $x > \mathbf{k}$, and thus, use a strictly positive interest rate r, not necessarily constant, but evolving with the highest velocity $r'(t) = c$? This choice cannot govern a viable evolution because the "metastate" $(x(t), r(t))$ will reach at some finite time $t^\star$ a state $(x(t^\star), r(t^\star))$ of the boundary of the graph of the regulation map R_c, which is a viability kernel (by construction). As soon as this boundary is crossed, it will violate the upper bound **c** at some finite time.

Hence, *we are forced to switch the interest rate, by an impulsive change of their velocity*. The Quincampoix *"semipermeable barrier"* property of the part of the boundary of a viability kernel in the interior $]\mathbf{k}, \mathbf{c}[\times\mathbb{R}$, dictates the choice $r'(t) = -c$ (see Theorem 10.5.19, p. 409, of [15, *Viability Theory. New Directions*]). Continuing with the velocity $r'(t) = -c$ of the interest rate, an analogous story can be told because sooner or later, this evolution will also reach the boundary of the graph of the critical map R_c, where t*he impulse change of the velocity of the interest rate to* $r'(t) = +c$ *is mandatory* for keeping $y(t)$ in the interval $[\mathbf{k}, \mathbf{c}]$.

In other words, imposing viability constraints and an inertia threshold, we cannot leave open the choice of the velocities $r'(t)$ as soon as we are compelled to choose $|r'(t)| = c$ on the critical zone, triggering in this way the viability cycle, made of four different phases: Exuberance, Sluggish Growth, Deflation, and Stagflation.

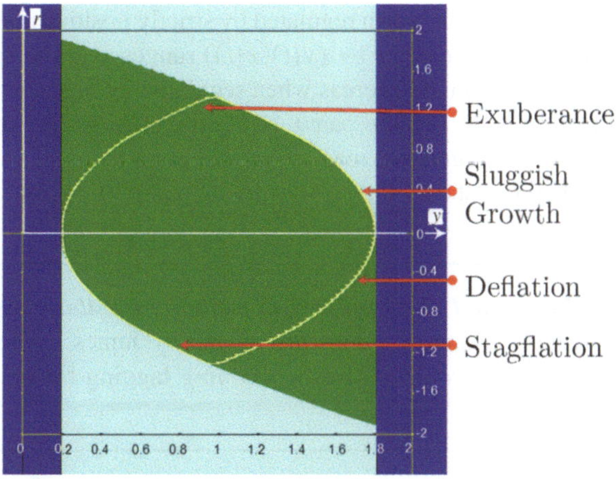

28 Viability Economic Crises. Software by Patrick Saint-Pierre.

1. **Exuberance.** This is the case when we start from $(\mathbf{k}, 0)$ with the highest velocity $r'(t) = +c$ of the interest rates. Since $y'(t) > 0$, both $y(t)$ and $r(t)$ increase, so that the "metastate" $(x(t), r(t))$ will reach at some finite time $t^\star$ a state $(x(t^\star), r(t^\star))$ of the boundary of the graph of the regulation map R_c.
2. **Sluggish Growth.** This is the case when we start from $(x(t^\star), r(t^\star))$ with the lowest velocity $r'(t) = -c$, dictated by the *barrier property* whenever $y(t) < \mathbf{c}$ before it reaches the upper bound $\mathbf{c}$ of the environment. Therefore the interest rate decreases while the value $y(t)$ continues to increase, slower and slower (*sluggish growth*). Note that during this time interval, the value-interest rate evolution satisfies the *Titanic effect*.

   Reaching $y(t^\star) = \mathbf{c}$, we could take $r(t^\star) = 0$ and remain at $\mathbf{c}$ forever. This is the case if we wish to describe the heavy evolution, but we choose to keep $r'(t) = -c$.
3. **Deflation.** This is the case when we start from $(\mathbf{c}, 0)$ with the lowest velocity $r'(t) = -c$. It is the symmetric phase to exuberance, when both the value and the interest rate decrease until $(x(t), r(t))$ reaches at some finite time $t^{\star\star}$ a state $(x(t^{\star\star}), r(t^{\star\star}))$ of the boundary of the graph of the regulation map R_c.
4. **Stagflation.** This is the case when we start from $(x(t^{\star\star}), r(t^{\star\star}))$ with the highest velocity $r'(t) = +c$. Then the interest rate $r(t)$ increases (inflation) but the value still decreases strictly (economic slow down). It is the symmetric counterpart of the sluggish growth phases, during which the *Titanic effect* takes place until the evolution reaches the equilibrium $(\mathbf{k}, 0)$. This phase of the economic cycle corresponds more or less to what economists qualify by stagflation, although the evolution does not stagnate, but, worse, decreases.

This ends the *viability cycle*: when regulated by strictly positive interest rates, the value goes from **k** to **c** and the pair $t \mapsto (y(t), r(t))$ ranges over the upper bound of the graph of the regulation map, whereas when regulated by negative interest rates, the value goes from **c** to **k**, and the pair $t \mapsto (y(t), r(t))$ ranges over the lower bound of the graph of the regulation map.

What we described is just a *hysteresis loop* by the viability oscillator which is an example of an *hysteron*:

29 ***[Hysteresis Loops and Hysterons]*** *This viability oscillator is an example of* hysteron *producing* hysteresis loops *discovered by* James Ewing, *a Scottish physicist who coined the word* hysteresis, *meaning* lagging behind *in classical Greek.*

This also illustrates cyclic ecological evolutions in population dynamics, as well as in many other fields (genetics, neurosciences, biology, etc.).

Chapter 5
Evolutions and Their Temporal Windows

The fateful day is near, and comes like a thief,
Prior to our mistakes heaven abandon us.
Take this moment of grace that he gives us.
Time flies and draws us with it.
The moment in which I am speaking is already far from me.

Nicolas Boileau, Epistles (1669–1695)
Epistle III to M. Arnaud

We did not respect in the illustrations presented in Chaps. 2, p. 31, and 3, p. 47, the venerable tradition inherited from physical sciences studying mainly evolutions on *given* temporal windows $[0, T]$ or $[0, +\infty[$ governed by differential systems starting at the origin 0 of the chronological time at specific *initial conditions* up to the future end.

As far as life sciences systems are concerned, the future is unknown and the best we can do is to *understand the past on temporal windows* $[T - \Omega, T]$ the aperture of which are part of the solution, as we did when we observed the illustrative examples. In this chapter, we justify this choice and propose some suggestions concerning temporal windows and the computation of aperture of temporal windows under which evolutions satisfy required properties.

5.1 Evolutions and Their Temporal Windows

We begin with a series of definitions leading to the concept of evolutions defined on temporal windows, later on associated with *duration functions* doomed to chaperon the evolutions of the state over which they evolve.

J.-P. Aubin, *Time and Money*, Lecture Notes in Economics and Mathematical Systems 670, DOI 10.1007/978-3-319-00005-3_5,

1. **The Spatial Metaphor** of time associates with the fleeting[1] *present time* t its position in the (geometric) real line $t \in \mathbb{R}$. Although this choice is debatable and has been intensively debated over centuries, we accept it to represent time as a real number.
2. **Temporal Windows.** We do not need the concept of an origin of time for defining a temporal window $[T - \Omega, T]$. It is replaced by the requirement that its *aperture* $\Omega \in \mathbb{R}_+$ is positive and thus, that 0 is the origin of the "duration space" of apertures.

   Fixing *an origin* $\mathbf{O}$ *of "chronological time"* $t \in \mathbb{R}$ amount to identify future times T (often regarded as horizons, exercise time in finance, etc.) with apertures $\Omega := T - \mathbf{O}$ and study what happens in *auspicious* temporal window $[\mathbf{O}, T]$, *after* its beginning.

   This may be too strong an assumption. We suggest to consider instead what happened in *propitious* temporal windows $[T - \Omega, T]$ *before* its end T. This end T may evolve as a fleeting present whereas an arbitrarily fixed origin separates past and future forever.[2]
3. **Evolutions.** This allows us to study evolutions on temporal windows when *both time* T *and* $\Omega \geq 0$ *evolve independently*, by $(T, \Omega) \in \mathbb{R} \times \mathbb{R}_+$[3]:

Definition 5.1.1 (Evolutions). Let us consider a set X (without identified algebraic or topological structure for the time).

An *evolution* $x(\cdot) : t \mapsto x(t) \in X$ in the set X is any single-valued map $t \in [T - \Omega, T] \mapsto x(t) \in X$.

It depends both on chronological time t **and** on the temporal window $[T - \Omega, T]$ (called the domain of the evolution) described by $(T, \Omega) \in \mathbb{R} \times \mathbb{R}_+$, i.e., it is a function of the three variables (T, Ω, t) where $t \in [T - \Omega, T]$.

Recall that the *trajectory* of an evolution $x(\cdot)$ is the subset $\{x(t)\}_{t \in [T-\Omega,T]}$.

Remark. To be rigorous would require to denote an evolution $x(\cdot)$ for any time t in the temporal window $[T - \Omega, T]$ as a set-valued map $(T, \Omega, t) \rightsquigarrow \tilde{x}(T, \Omega, t)$ defined by

---

[1] Adjective coined in 1882 by *Robert Kelly*, the anonymous author of [93, *The Alternative: a Study in Psychology*], which survived only because *William James* popularized it in [91, *Principles of Psychology*] (see [2, Andersen & Grush]). It has and will run much ink with the progress of cognitive science.

[2] *"What was God doing before He created the Heavens and the Earth?"* asked *Augustine of Hippo* in his confessions. Is his eternity only forward in time and not backward? Introducing the concepts of temporal windows and duration function bypasses the question of origin of time.

[3] As *Dobbs* proposed in 1951, instead of the usual representation by a point of the line. This philosopher was motivated by the mathematics of his time, and chose metaphors based on quantum physics and the uncertainty principle ([58, 59, Dobbs], [79, Grünbaum], [39, Broad]) to give some "thickness" to "transitory time" by introducing a second component of the time, the "extensional time" to give a formal sense the specious present. To our knowledge, these attempts were not pursued. I thank *Jean-Philippe Terreaux* for having mentioned this reference.

$$\tilde{x}(T, \Omega, t) := \begin{cases} \{x(t)\} \text{ if } t \in [T - \Omega, T] \\ \emptyset \qquad \text{otherwise} \end{cases} \tag{5.1}$$

We shall say that t is an *instant* if it is identified with the triple $(t, 0, t)$ of *aperture* 0 ($t \in [t, t]$). Regarding the temporal window as the set-valued map $t \mapsto \mathbb{I}(T, \Omega) := [T - \Omega, T]$, of which it is the *domain*, the triples (T, Ω, t) range over the graph Graph($\mathbb{I}$) $\subset \mathbb{R} \times \mathbb{R}_+ \times \mathbb{R}$ of $\mathbb{I}$. "Classical evolutions" are thus defined on the subset $\mathscr{T} := \{(t, 0, t)_{t \in \mathbb{R}^3}\}$ which do not specify the temporal window on which they are defined. ■

4. **Chronological Time: The Numéraire of Evolutions.** The *chronological time* is by definition the specific evolution $t \in \mathbb{R} \mapsto \tau(t) := t \in \mathbb{R}$ having a velocity equal to 1, vanishing at 0 regarded as the origin $\mathbf{O} = 0$. It can be regarded as *a "numéraire" of all evolutions*, allowing them to be compared by "comparing" each of them to the chronological time which they use in their definition. This is comparable with the "numéraire" in economics, used for comparing commodities by comparing the value of each of them with the value of the numéraire (see more details in Sect. 8.2, p. 601 of [12, *La mort du devin, l'émergence du démiurge*]). Indeed, chronological time can be interpreted as the *chaperon* of any evolution which, as it is defined in Definition 5.1.1, p. 68, can be reformulated as the pair of evolutions

$$\begin{cases} (i) \ t \mapsto \tau(t) \ = t \\ (ii) \ t \mapsto x(t) \end{cases} \tag{5.2}$$

The evolution of the state is de facto compared to the evolution of the chronological time.

5. **Calendar Durations.** There are other examples of evolutions $t \in \mathbb{R} \mapsto d(t) \in \mathbb{R}$ with velocity $d'(t) = 1$ which do not depend upon the choice of the time origin. These evolutions are called *calendar durations*. Their constant velocities equal to 1 by definition describe mathematically the concept of *time arrow* implying the irreversibility of time.

   An example of calendar evolution is the age $t \in [T - D, T] \mapsto d(t) := \max(0, t - (T - D))$ associated with the time–age pair (T, D) defined on the temporal window [T-D,T]. The values $d(t) \in \mathbb{R}_+$ being nonnegative, they range over the *duration space* $\mathbb{R}_+$ *which has an origin* (age equal to 0 at birth), the choice of it being no longer debatable. It increases from $d(T - D) = 0$ to $d(T) := D$, where $T - D \in \mathbb{R}$ is the date of birth.

   In mathematical demography and population dynamics, McKendrick and Hamilton–Jacobi–McKendrick partial differential equations provide the state of a population depending on chronological time T and on age D (they are "structured" by age).

   Age-chaperoned evolutions are governed by differential equations $x'(t) = F(t, t - (T - D), x(t))$ underlying "age-structured" McKendrick partial differential equations,[4] which could be rewritten as

$$(d'_{\text{chronological}}(t), d'_{\text{calendar}}(t), x'(t)) = \{1\} \times \{1\} \times F(d_{\text{chronological}}(t), d_{\text{calendar}}(t), x(t))$$

They are chaperoned by two calendar durations: chronological time and age.

6. **Duration Functions.** The following question pops up: *why do we have to restrict to* prescribed *"chronological time" $t \in [0,T] \mapsto \max(0,t) \geq 0$ and calendar age $t \in [T-D,T] \mapsto \max(0, t-(T-D)) \geq 0$, both with constant velocity equal to 1? Why not should we choose and regulate increasing duration functions $t \mapsto d(t)$ chaperoning the evolution of states for determining apertures $\Omega(d(\cdot))$ of temporal windows over which the evolutions satisfy required properties?*

   I suggest to answer these questions by introducing duration functions in Sect. 5.2, p. 70. We refer to the book [103, *Biological Time, Symmetries and Singularities*] by *Giuseppe Longo* and by *Maël Montévil* for more information on this topic.

5.2 Duration Functions

This section continues Sect. 5.1, p. 67, which suggested to derive temporal windows from duration functions.[5]

Definition 5.2.1 (Duration functions).

1. A *duration function* is any (differentiable) evolution $d(\cdot) : t \in \mathbf{R} \mapsto d(t) := d_{(T,D)}(t) \in \mathbb{R}_+$ (regarded as the duration space associated with) a *time–duration pair* (T, D) and satisfying

   a. $d(\cdot)$ is *a nondecreasing nonnegative function*;
   b. *ending at D at T*: $d(T) = D$.

[4]The age-structured standard approach starts with the establishment of the McKendrick partial differential equation relating the population and its partial derivatives with respect to time and age. They do not involve explicitly calendar evolutions, except in some underlying proofs. Age-structured partial differential equations involving both time and age have been studied (see, among an extensive literature, [3, Anita], [14, Aubin], [16, Aubin, Bonneuil & Maurin], [90, Iannelli],[94, *Applied Mathematical Demography*], [95, Keyfitz N. & Keyfitz B.], [122, Von Foerster], [124, Webb], etc.).

[5]That may be called *durances*: *Efim Galperin* taught me that *duration* is, in English, constant, so that variable duration is an oxymoron. He also told me that there exists a synonym in English, *durance*, which could be used for duration functions. It is derived from Old French (see the definition: *DURABLETË; durance s. f. durée, perpétuité. Prov. durabletât. It. durabilità* of [36, *Glossaire de la langue d'oïl,(XIième-XIVième siècles)*] by *Alphonse Bos* and *durableté, -ance, n. f. (XIIième-XVième) : 1. Durée : " la nature divine laquelle ... donne durance, gloire, honneur " (ième siècle, Jean Le Bel); 2. Endurance. Étymologie Durable, Durer* of [77, *Dictionnaire d'ancien français. MoyenÂge et Renaissance*] by *Robert Grandsaignes d'Hauterive*), that he suggested to use in English and to resuscitate in French.

2. The **aperture of the temporal window** $\Omega(d(\cdot))$ of the duration function $d(\cdot)$ is defined by

$$\Omega(d(\cdot)) := \Omega_{T,D}(d_{T,D}(\cdot)) := \inf_{\{\omega \geq 0 \text{ such that } d_{T,D}(T-\omega)=0\}} \omega \in [0, +\infty]. \tag{5.3}$$

3. The **temporal window** of the duration function $d(\cdot)$ is the interval $[T - \Omega(d_{T,D}(\cdot)), T]$, its *beginning* being equal to $T - \Omega(d_{T,D}(\cdot))$ and its *ending time* being equal to T;
4. The **duration fluidity** is its velocity $\varphi(t) := d'(t)$, which is *positive or equal to* 0 since duration functions do not decrease with time, but no longer necessarily equal to 1, as for calendar duration functions.

For simplicity, we denote by $d(\cdot) := d_{(T,D)}(\cdot) \mapsto \Omega(d(\cdot)) := \Omega_{T,D}(d_{T,D}(\cdot))$ the *aperture functional* which is defined on the set of duration functions ending at D at time T, and thus, indexed by the time–duration pair (T, D).

Duration functions, including the chronological time, being increasing, are *irreversible*. Only evolutions governed by differential equations involving *derivatives of even order are reversible* in (chronological time), since inverting the time does not change the sign of those derivatives, in particular, the sign of the acceleration. Some evolutions may be "reversible" in time, but chronological time and duration functions are (assumed) to be irreversible.[6]

Since a duration function d is a nondecreasing nonnegative function, it satisfies $d(t) = \max(0, d(t))$. This remark allows us to associate with any nondecreasing function $e(\cdot)$ a duration function $d(\cdot)$ by setting $d(t) := \max(0, e(t))$.

We distinguish temporal windows

1. *with free aperture* $\Omega(d(\cdot))$ (computed through the duration function);
2. *with prescribed aperture* $\Omega = \Omega(d_{(\Omega,T,D)}(\cdot))$ where $\Omega := \Omega(d_{(\Omega,T,D)}(\cdot))$, constraining the duration function $d_{(\Omega,T,D)}(\cdot)$ to depend also on the required aperture Ω.

Every duration function $d_{(T,D)}(\cdot)$ can be written in terms of the time–duration pair (T, D) and its fluidity by: $\forall\, t \leq T,\ d_{(T,D)}(t) = \max\left(0, D - \int_t^T d'_{(T,D)}(\tau)d\tau\right)$. Among them, duration functions with constant fluidity $\varphi \in [0, +\infty[$ defined by: $\forall\, t \in \left[\frac{\varphi T - D}{\varphi}, T\right]$, $d_\varphi(t) = \max(0, D - \varphi(T - t))$, the aperture $\Omega_{(T,D)}(d_\varphi(\cdot)) = \frac{D}{\varphi}$ of which *decreases as the constant fluidity increases*.

---

[6]Some physicists would suggest that "entropy" could be a candidate to the role of duration function. This is not the place to open this voluminous Pandora Box.

1. The *constant duration* $d_0(\cdot)$ defined by $d_0(t) = D, t \leq T$, with zero fluidity and infinite aperture $\Omega_{T,D}(d_0(\cdot)) = \infty$;
2. The *calendar duration* $d_1(\cdot)$ defined by $d_1(t) := \max(0, t - (T - D))$ on the temporal window $[T - D, D]$ of aperture equal to D;
3. The *constant fluidity duration* φ are equal to

$$d_\varphi(t) = \max(0, D - \varphi(T - t)) \tag{5.4}$$

the aperture of which is $\Omega_{(T,D)}(d_\varphi(\cdot)) = \dfrac{D}{\varphi}$;
4. The *instantaneous duration*[7] $d_\infty(\cdot)$ (defined by $d_\infty(t) = 0$ if $t < T$ and by $d_\infty(t) = [0, D]$ if $t = T$), with infinite fluidity and zero aperture.

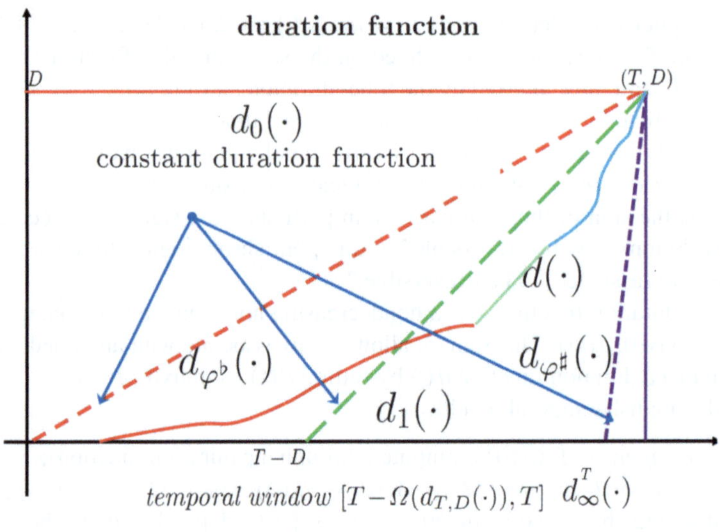

30 (Duration Functions) *This diagram shows the graphs of several duration functions associated with calendar pairs (T, D). Among them are the constant duration function $d_0(\cdot)$, the calendar time $d_1(\cdot)$, two constant fluidity duration functions $d_{\varphi^\flat}(\cdot)$ and $d_{\varphi^\sharp}(\cdot)$ and duration function $d(\cdot)$ with variable fluidity.*

[7] The instantaneous duration function $d_{+\infty}(\cdot)$ is no longer a single-valued map, but either the set-valued map equal to 0 for $t < T$ and $[0, D]$ for $t = T$, the graphical derive of which at $(T, 0)$ is infinite or else, the *Heavyside* distribution, the derivative of which is the Dirac measure at T. It is not a member of $\mathscr{A}_{[\varphi^\flat, \varphi^\sharp]}(T, D)$ and we shall not use them in this study.

Definition 5.2.2 (Duration Functions with Bounded Fluidity). When $0 \leq \varphi^\flat \leq \varphi^\sharp < +\infty$, we denote by

$$\mathscr{A}_{[\varphi^\flat,\varphi^\sharp]}(T,D) \ := \ \{d(\cdot) \text{ such that } d'(\cdot) \in [\varphi^\flat, \varphi^\sharp]\} \tag{5.5}$$

the space of *duration functions with bounded fluidity* $d(\cdot)$ between $\varphi^\flat$ and $\varphi^\sharp$.

They satisfy inequalities

$$d_{\varphi^\sharp}(t) \ := \ \max(0, D-\varphi^\sharp(T-t)) \ \leq \ d_{(T,D)}(t) \ \leq \ \max(0, D-\varphi^\flat(T-t)) \ =: \ d_{\varphi^\flat}(t) \tag{5.6}$$

There is much concern these days with the pressure of productivity, speed of communications, ubiquity associated with immediacy, high frequency transactions, usually described as the *time acceleration*. Mathematically, it can be defined as *the second derivative of the duration function*. A *duration function with prescribed fluidity* $d(\cdot) := d_{(T,D,\varphi)}(t)$ is a *duration function* ending at T with a duration $d(T) = D$ and with a ending fluidity $d'(T) = \varphi$, which can be written as

$$\forall\, t \ \leq \ T, \ \ d_{(T,D,\varphi)}(t) \ = \ D - \varphi(T-t) + \int_t^T (T-\tau)d''_{(T,D,\varphi)}(\tau)d\tau \tag{5.7}$$

Among them, duration functions $d_{(\varphi,\gamma)}(\cdot)$ with constant accelerations γ and terminal *fluidity* φ are parabolas of the form

$$d_{(\varphi,\gamma)}(t) \ := \ \max\left(0, D-\varphi(T-t)+\gamma\frac{(T-t)^2}{2}\right)$$

convex for positive accelerations $\gamma \geq 0$ and concave for negative accelerations $\gamma = -|\gamma|$, satisfying

$$\begin{cases} (i) \ \ d(T) = D \\ (ii) \ \ d'_{(\varphi,\gamma)}(t) \ = \ \varphi - \gamma(T-t) \text{ and thus, } d'_{(\varphi,\gamma)}(T) \ = \ \varphi \\ (iii) \ d''_{(\varphi,\gamma)}(t) \ = \ \gamma \end{cases}$$

The aperture of $d_{(\varphi,\gamma)}(\cdot)$ is equal to

$$\Omega(d_{(\varphi,\gamma)}(\cdot)) \ = \ \begin{cases} \dfrac{\sqrt{\varphi^2+2|\gamma|D}-\varphi}{|\gamma|} & \text{if } \gamma = -|\gamma| \ \leq \ 0 \\ \dfrac{\varphi+\sqrt{\varphi^2-2\gamma D}}{\gamma} & \text{if } \gamma \ \in \ \left[0, \dfrac{\varphi^2}{2D}\right] \end{cases} \tag{5.8}$$

Vectorial Duration Functions and Scheduling

We have also to consider the case when the evolution of the state $x(\cdot) := (x_i(\cdot))_{i=1,\ldots,n} \in \prod_{i=1}^{n} X_i$ ranges over a product of state spaces. It may be judicious to chaperon each component $x_i(\cdot)$ with a specific duration function $d_i(\cdot)$ determining the temporal window $[T - \Omega(d_i(\cdot)), T]$ during which this components evolves, and thus the evolution $x(\cdot) := (x_i(\cdot))_{i=1,\ldots,n} \in \prod_{i=1}^{n} X_i$ by a (vectorial) duration function $d(\cdot) := (d_i(\cdot))_{i=1,\ldots,n} \in \mathbb{R}^n_+$, regarded as the n-dimensional space. We set

$$\vec{\Omega}(d(\cdot)) := \Omega_{T,D}(d(\cdot)) := \inf_{\{\omega \geq 0 \text{ such that } d(T-\omega)=0\}} \omega \in [0, +\infty] \tag{5.9}$$

This implies that for any component $i = 1, \ldots, n$, $d_i \vec{\Omega}(d(\cdot)) = 0$, so that

$$\vec{\Omega}(d(\cdot)) := \max_{i=1,\ldots,n} \Omega(d_i(\cdot))$$

This induces a *ranking* on the "component durations", and the *scheduling* of the temporal widows $[T - \Omega(d_i(\cdot)), T] \subset [T - \vec{\Omega}(d(\cdot)), T]$:

$$[T - \vec{\Omega}(d(\cdot)), T] = \bigcup_{i=1}^{n} [T - \Omega(d_i(\cdot)), T] \tag{5.10}$$

5.3 Maupertuis' Pleasure

The concept of duration was already used by *Maupertuis*. He invented not only the *principle of least action*, but adapted it for defining pleasure in his [110, *Essay de philosophie morale*][8]:

> **31 (Maupertuis' Definition of Pleasure)** "The estimation of happy and unhappy moments is the *product* of the *intensity* of the pleasure or pain by the *duration*".

However, *Maupertuis* did not elaborate this concept in his short essay, and did not study kinds of potential functions $U : d \mapsto U(d)$ on durations, interpreted as a *"perception of time,"* rather, of duration.

[8] Among so many Maupertuis contributions, see [106–109, Maupertuis].

Assuming that the perception of time function U is differentiable, its derivative $e := U'(d)$ could regarded as a candidate for the role of "intensity of pleasure." I suggest to interpret this derivative as an *"urge"* to act, the analogous of a force in mechanics (or a price in economics). Whenever the fluidity of time is equal to 1, then the velocity of evolution $t \mapsto U(d(t))$ is equal to $U'(d(t))$, which is the "estimation of happy moments," the analogous of "work" in mechanics.

What happens if the duration has a variable fluidity?

When the perception of time $t \mapsto U(d(t))$ evolves, its derivative $\frac{dU(d(t))}{dt} = U'(d(t))d'(t)$ is the product of the urge $U'(d(t))$ and of the fluidity $d'(t)$. It is regarded as the *"impetuosity"* along the time, the analogous of "power" in mechanics (or of a transaction value in economics). This power could be the candidate for the underlying concept of happiness in Maupertuis' world.

The urge can also evolve, and its velocity could be regarded as a *"haste"*. The product of a duration by an urge and by a haste could be interpreted as *"precipitation"* and *"impulsiveness"* respectively.

This looks as a zoological attempt to coin names in a ménagerie of concepts. But since derivatives of durations and fluidities are involved in a dynamic approach of time, it may be relevant to use synonyms of the same idea of the vernacular language and to give them precise meanings, useful in a mathematical study. They are summarized in the following diagram:

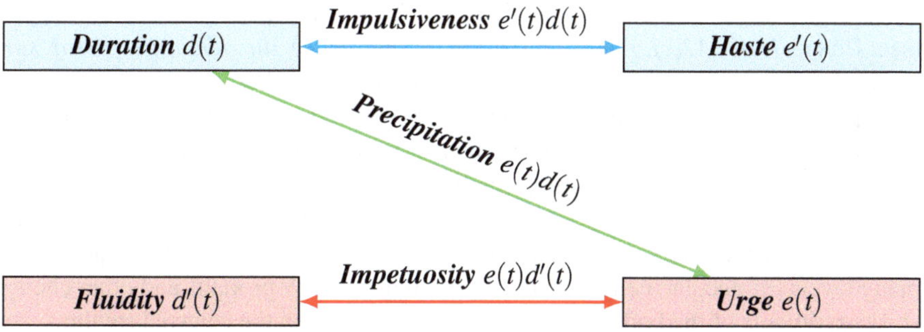

These concepts have been introduced by Maupertuis and his successors, *Leonhard Euler*, *Joseph*, *Lagrange*, *William Hamilton*, *Carl Jacobi* in the framework of the calculus of variations, etc. As well as in economics, as we shall see in Sect. 7.1, p. 105 when durations chaperon the evolution of commodities and prices.

Who could be a better mentor than Maupertuis*?*

5.4 Giving Duration Time

The favorite adage of *François Mitterrand* was "giving time time" (*donner du temps au temps*). Splitting the polysemous concept into two meanings in this study among many other ones, chronological time and duration functions, we proposed to make more precise this adage by replacing it by *donner du temps à la durée*, which means "giving duration time."

Is it reasonable to recall what *Augustine of Hippo* wrote in Book XI of his *Confessions*: *"What, then, is time? If no one ask of me, I know; if I wish to explain to him who asks, I know not. Yet I say with confidence, that I know that if nothing passed away, there would not be past time; and if nothing were coming, there would not be future time; and if nothing were, there would not be present time. Those two times, therefore, past and future, how are they, when even the past now is not; and the future is not as yet?"*

Since then, at least, so many philosophers and scientist wrote on this topic that it is tempting to say that everything was thought over and told and that it may be pretentious to have something to add.

However, at least, it is possible to reduce the polysemy of time. This is the reason why, motivated both by mathematics and economics dealing with the concept of value, we made the above suggestions, which probably have already been proposed under other names and in other fields of knowledge.

I just mention briefly that the "phylogenesis" of time has "taken all its time" to synchronize the "perceptions" of living things by their biological clocks with those that the facetious *celestial mechanics* "turning ellipse" our planet instead of "turning round" by bending it over its ecliptic. However, the strange cyclical nature of celestial bodies is the source of the perception of "duration."

Living organisms record (store) in their "cognitive system" perceptions *of* the environment together with their actions *on* the environment which they trigger (see Chap. 7 of [12, *La mort du devin, l'émergence du démiurge. Essai sur la contingence, la viabilité et l'inertie des systèmes*]). These joint evolutions of perception–actions, defining *behaviors* of the organism, are then maintained as cyclic evolutions of proteins in the organism, hormones as well as neurotransmitters among other molecules, emitting, sending, and receipting them. These cyclic perceptions are tuned with the ephemeridal cycles (circadian, monthly, seasonal, annual, etc.). Since *Jean-Jacques de Mairan* in 1729, chronobiology is studying at many levels (molecular, cellular, etc.) the circuits of many proteins and other chemical constituents of the bodies coming into resonance with the celestial rhythms.

Memory plays a crucial role in recording those perception–action pairs during temporal windows of varying aperture, by using the Lorenzian imprinting mechanism (originally described by *Douglas Spalding* in the nineteenth century and later by *Oskar Heinroth*, Lorenz's mentor): once recognized by a more or less similar perception recorded during a previous temporal window, it triggers one of the actions associated with this previous perception. In our brains, the notion of repetition of "equivalent" perception–action pairs was easier to detect than that of their "succession." This requires indeed a notion of order stating that a pair to perception–action is "before" another, as a step before the other one on a path: the *spatial metaphor*, where dates were part of a line, slowly superimposed to the *cyclic perception*.

"Time" then "emerges" in the cognitive system of an organism as soon as it recognizes, remembers, and "orders" the stored and present perception–action pairs. A "scheduling operation" requires a faculty of *classifying* perception–action pairs, of *comparing* and *ordering* them, and therefore provides the concepts of *repetition* and *succession*. Both time's arrows and time's cycles, the theme of one of the books, [76, *Time's Arrow, Time's Cycle*], by *Stephen J. Gould*, draw together a kind of helix, where perception–action pairs are stored, classified as equivalent, or ordered by biological mechanisms that are still far from being elucidated. A *helicoidally time*, somehow.

The chronological time was measured by the ephemerides through gnomons and sundials, and now, by clocks,[9] whereas the duration was obtained by clepsydra, and, since the tenth century, by hourglasses, or by the difference between two chronological times.

32 (Measuring the Time and Duration) Chronological time is measured by clocks, such as the 1410 magnificent astronomical clock of Prague. The duration can be measured by the time that elapses from the bottleneck of the hourglass, such as that of *memento mori* "Vanity" by Philippe de Champaigne (Le Mans Museum), glorifying the beautiful nature, symbolized by the tulip. Both the Prague clock and this picture display a skull for depicting the end of our personal temporal windows. For computing longitude (depending on the time difference observed with the sun), duration was measured by hourglasses every half an hour. Sleep hungry "eating sand" sailors turned the bulb of the hourglass before it was completely empty, risking their lives at sea! Monks themselves, shortening still variables hours to shorten their fasts, risked also their lives, but in the hereafter. In some monasteries, the moon was used to estimate the duration by the number of windows lit during its path to determine "temporal windows": *luna se fundebat per fenestras*.

[9]Without entering philosophical considerations, following the paradoxes of *Zeno of Elea*, we may observe, as *Bergson* pointed out in [32, 33, *Creative Evolution*], that the perception of a continuous evolution of time is fallacious, that time reduces to a "dust of instants" (poussière d'instants). Measuring chronological time would require to stop it for taking a (static) measure. But measures do not need to be static: assuming that the time is a numéraire of evolution, it can be dynamically measured by clocks *"reproducing"* evolutions with velocity equal to 1.

5.5 Evolutionary Systems

We denote by $\mathscr{C}(\mathbb{R}, X)$ the space of continuous evolutions. We can always define the operations of translation and concatenation, whatever the algebraic or topological structures on the set X (however, in the examples of this study, X is a finite dimensional vector space).

Definition 5.5.1 (Translations and Concatenations).

1. **Translation.** Let $x(\cdot) : \mathbb{R} \mapsto X$ be an evolution defined on the temporal window $[T - \Omega, T]$. For all $\sigma \geq 0$, the translation (to the left) $\kappa(-\sigma)(x(\cdot))$ of the evolution $x(\cdot)$ is defined by $\kappa(-\sigma)(x(\cdot))(t) := x(t + \sigma)$ on the (translated to the left) temporal window $[T - \Omega - \sigma, T - \sigma]$ and the translation (to the right) is defined by $\kappa(+\sigma)(x(\cdot))(t) := x(t - \sigma)$ on the temporal window $[T - \Omega + \sigma, T + \sigma]$;
2. **(Time) Reverser.** The *(time) reverser* ρ_T associates with any evolution $x(\cdot)$ defined on a temporal window $[T - \Omega, T]$ the evolution $\rho_T(x(\cdot))$ defined on the temporal window $[0, \Omega]$ by
$$\forall\, t \in [0, \Omega],\ \ \rho_T(x(\cdot))(t) \ := \ x(T - t) \tag{5.11}$$
The reverser maps any evolution $\overleftarrow{x}(\cdot)$ defined on the temporal window $[0, \Omega]$ to the evolution defined on the temporal window $[T - \Omega, T]$ by $\rho_T(\overleftarrow{x}(\cdot))(t) = x(T - t)$ (see Fig. 16, p. 43). It is idempotent in the sense that
$$\forall\, x(\cdot),\ \ \rho_T(\rho_T(x(\cdot))) \ = \ x(\cdot) \tag{5.12}$$
Whenever there is no ambiguity, we set
$$\overleftarrow{x}(\cdot) \ := \ \rho_T(x(\cdot)) \tag{5.13}$$
for simplifying notations.
3. **Concatenation.** Let $x(\cdot) : \mathbb{R}_+ \mapsto X$ and $y(\cdot) : \mathbb{R}_+ \mapsto X$ be two evolutions. For all $T \geq 0$, the *concatenation* at time T $(x(\cdot) \diamond_T y(\cdot))(\cdot)$ of the evolution $x(\cdot)$ defined on the temporal window $[T - \Omega, T]$ and $y(\cdot)$ defined on $[0, S]$ is the evolution defined by
$$(x(\cdot) \diamond_T y(\cdot))(t) \ := \begin{cases} x(t) & \text{if } t \in [0, T] \\ \kappa(+T)(y(\cdot))(t) := y(t - T) & \text{if } t \geq T \end{cases} \tag{5.14}$$
It is single-valued if and only if $x(T) = y(0)$.

The translations satisfy the property

$$\forall\, \tau, \sigma \in \mathbb{R}, \;\; \kappa(\tau+\sigma) \;=\; \kappa(\tau)\circ\kappa(\sigma) \;=\; \kappa(\sigma)\circ\kappa(\tau) \tag{5.15}$$

Definition 5.5.2 (Differential Inclusions and their Evolutionary Systems). In this study, the evolution are governed by differential equations $x'(t) = f(t, x(t))$, but, most often, by differential inclusions $x'(t) \in F(t, x(t))$ (see [21, *Differential Inclusions. Set-Valued Maps and Differential Inclusions*]).

1. With any terminal time T and any terminal state x, we denote by $\mathscr{A}(T,x)$ the (possibly empty) set of evolutions $x(\cdot)$ governed by differential inclusion $x'(t) \in F(t, x(t))$ *arriving* at x at terminal time T;
2. We denote by $\mathscr{S}(\Omega, x)$ the set of reverse evolutions governed by differential inclusion $x'(t) \in F(T-t, x(t))$ *starting* at x at initial duration 0 up to aperture Ω.

We shall say that the evolutionary system is *deterministic* on a subset $K \subset X$ if for any T, the map $x \mapsto \mathscr{A}(T,x)$ is single-valued, or, equivalently, if its reverse system $x \mapsto \mathscr{S}(T,x)$ is single-valued.

We mention that *differential equations can generate non deterministic systems and differential inclusions can generate deterministic ones* in the sense of this definition (see [15, *Viability Theory, New Directions*]). This definition can be felt ambiguous, since quite often, differential equations where velocities associated with states being univoque are described as deterministic and velocities associated with differential inclusions are felt as non-deterministic. What matters, in out opinion, is the univocity of the evolutions governed by the evolutionary system and not the univocity of velocities defining the system.

Lemma 5.5.3 (Properties of Evolutionary Systems). *Let us consider a differential inclusion $x'(t) \in F(t, x(t))$.*

An evolution $x(\cdot)$ arriving at x at terminal time T belongs to $\mathscr{A}(T,x)$ if and only if the reverse evolution $\overleftarrow{x}(\cdot) := \rho_T(x(\cdot))$ starting at x at initial duration 0 *belongs to $\mathscr{S}(T,x)$.*

An evolution $x(\cdot)$ is viable in a viability tube $K(\cdot)$ on the interval $[T-\Omega, T]$ *if and only if its reverse $\overleftarrow{x}(\cdot) := \rho_T(x(\cdot))$ is viable in the reverse tube $\overleftarrow{K}(\cdot) := \rho_T(K(\cdot))$ on the interval* $[0, \Omega]$.

An evolution $x(\cdot)$ starts at $t^\star$ from a departure tube $C(\cdot)$ at time $t^\star \leq T$ if and only if its reverse evolution $\overleftarrow{x}(\cdot) := \rho_T(x(\cdot))$ captures the target tube $\overleftarrow{C}(\cdot) := \rho_T(C(\cdot))$ at $\Omega^\star := T - t^\star \geq 0$.

Differential inclusions have been introduced in economics in the study of planning procedures: See, among many references, [47, 48, Champsaur, Drèze & Henry], [49–54, Cornet], [61, Drèze G. & de la Vallée Poussin], [105, Malinvaud].

5.6 Anticipating from the Past or Predicting the Future?

Speaking of the future requires the knowledge of a *deterministic evolutionary system* producing a *unique* evolution defined on the temporal window.

Definition 5.6.1 (Deterministic Evolutionary Systems Predicting the Future). We shall say that a *deterministic evolutionary system predicts the future* on temporal windows of fixed aperture Ω if for all evolutions $x(\cdot)$ governed by the evolutionary system on a temporal window $[T - \Omega, T]$, every translation $\kappa(\tau)x(\cdot)$ on the temporal window $[T - \Omega + \tau, T + \tau]$ is also governed by this deterministic evolutionary system. In particular, the properties of one evolution governed on a temporal window are shared (or invariant) by the translated evolutions on their translated temporal windows.

For *engineering* sciences, where the deterministic evolutionary systems are built by human brains, as well as for many domains of physics (not all, however), it is reasonable to assume that the *"the future can be predicted."* We refer to Sect. 5.7, p. 81 for more details on *extrapolators* extending the concept of evolutionary system when the "inputs" are no longer initial states, but "histories."

Life sciences evolutionary systems, for which experimentation is not possible, since they are non-deterministic and, even worse, *irreversible* in the sense that they cannot be reconstructed, it may be wiser to understand and *explain* what *happened in the past for anticipating evolutions* instead of *forecasting* what will happen in the future: according to *Paul Valéry*, *"Forecasting is a dream from which reality wakes you up"* ("La prévision est un rêve dont la réalité vous tire").

Life sciences require *redundancy* of regulons (regulatory controls, evolving without the action of a consensual actor) chosen in a "contingent reservoir" for adapting the evolutions to viability constraints or/and reaching objectives. This justifies the introduction of differential inclusions. Moreover, this must be the case whatever the *"tyches"* (unpredictable perturbations or disturbances) evolving in a "tychastic reservoir," motivating the introduction of tychastic differential equations and inclusions, which are not taken into account here (see [17, *Tychastic Measure of Viability Risk. A Viabilist Portfolio Performance and Insurance Approach*]).

The requirements for forecasting the future are no longer satisfied. The problem is then to design a *regulation map* that *anticipates* what to do whenever one of these viability or capturability properties is about to be violated. The regulation map provides at least one regulon correcting the situation by restoring the viability on a temporal window of strictly positive aperture.

In a nutshell, *anticipation* replaces for life science evolutionary systems the concept of *prediction*. The problem is not to predict when a "catastrophe" will happen (for which only human brains are addicted), but to provide the regulation maps for restoring the viability when the catastrophe does happen (a strategy used by living organisms since they exist).

This (non-Popperian) viewpoint seems to be more adequate for studying life science systems, among which, economic systems, whereas the former classical approach is more relevant for systems of physical and engineering sciences abiding by the Popperian edicts.

5.7 Extrapolation by Historical Differential Inclusions

Regarding $T \geq 0$ as an *evolving present time*, we can regard the translation $\kappa(-T) : \mathscr{C}(-\infty, +\infty; X) \mapsto \mathscr{C}(-\infty, 0; X)$ as a *recording operator* and translation $\kappa(+T) : \mathscr{C}(-\infty, +\infty; X) \mapsto \mathscr{C}(0,_\infty ; X)$ as a *recalling operator* in the sense that

1. $\kappa(-T)(x(\cdot))_- \in \mathscr{C}(-\infty, 0; X)$ can be regarded as the history of the evolution up to time T of the evolution $x(\cdot)$;
2. $\kappa(+T)(x(\cdot))_+ \in \mathscr{C}(0, \infty; X)$ can be regarded as the future of the evolution from time T of the evolution $x(\cdot)$.

"Histories" are evolutions $\varphi \in \mathscr{C}(-\infty, 0; X)$ defined for negative times. The historical space $\mathscr{C}(-\infty, 0; X)$ is a "storage" space in which we place at each $T \geq 0$ any evolution $x(\cdot)$ defined on $[-\infty, T[$ up to time T thanks to the translation operator $\kappa(-T)$.

Operating on evolutions require a specific calculus of "Clio derivatives" of history-dependent maps.

For instance, let a history-dependent functional $\mathbf{v} : \varphi \in \mathscr{C}(-\infty, 0; X) \mapsto \mathbf{v}(\varphi) \in \mathbb{R}$. The addition operator $\varphi \mapsto \varphi + h\psi$ is replaced by the *concatenation operator* $\diamond_h$ associating with each history $\varphi \in \mathscr{C}(-\infty, 0; X)$ the function $\varphi \diamond_h \psi \in \mathscr{C}(-\infty, 0; \mathbf{R}^\ell)$ defined by

$$(\varphi \diamond_h \psi)(\tau) := \begin{cases} \varphi(\tau + h) & \text{if } \tau \in]-\infty, -h] \\ \varphi(0) + \psi(\tau + h) & \text{if } \tau \in [-h, 0] \end{cases}$$

This allows us to define the concept of *Clio*[10] *derivatives* by taking the limits of "differential quotients"

$$\nabla_h \mathbf{v}(\varphi)(\psi) := \frac{\mathbf{v}((\varphi \diamond_h \psi)) - \mathbf{v}(\varphi)}{h} \in X$$

for obtaining

$$D\mathbf{v}(\varphi)(\psi) := \lim_{h \to 0+} \nabla_h \mathbf{v}(\varphi)(\psi) \in X$$

The problems studied below could and should be adapted to historical differential inclusions, which constitute the natural framework for studying evolutions in life sciences.

[10] Clio, muse of history, was born as the other muses out of the love between Zeus and Mnenosyne, Goddess of memory.

Hence, instead of studying evolutions $t \mapsto x(t) \in X$, we associate evolutions $t \mapsto (\kappa(-t)x(\cdot)) \in \mathscr{C}(-\infty, 0; X)$ in the historical space. We can extend the concept of evolutionary system to the concept of extrapolator system generated by history-dependent (or path-dependent, memory-dependent, functional) differential inclusions associating with the history of the evolution up to each time t a set of velocities: histories are the inputs of history-dependent differential inclusions

$$x'(t) \in F(\kappa(-t)x(\cdot)) \tag{5.16}$$

where $F : \mathscr{C}(-\infty, 0; X) \rightsquigarrow \mathbb{R}^\ell$ is a set-valued map defining the dynamics of history-dependent differential inclusion (see [80–82, Haddad] and [25, 26, Aubin & Haddad]).

One can also use history-dependent differential equations or inclusions depending on functionals on past evolutions, such as their derivatives up to a given order m:

$$x'(t) \in F\left((D^p(\kappa(-t)x(\cdot)))_{|p|\leq m}\right) \tag{5.17}$$

in order to take into account not only the history of an evolution, but its "trends."

For instance, these history-dependent differential inclusions have been used for extrapolating the asset prices.

The *VIMADES Extrapolator* (based on Laurent Schwartz distributions) is an example of history-dependent differential inclusion by extrapolating each history-dependent al (past) evolutions of upper bounds (*HIGH*) and lower bounds (*LOW*) of the underlying prices provided by brokerage firms (see [19, Aubin, Chen, Dordan & Saint-Pierre], [20, Aubin, Chen, Dordan, Faleh, Lezan & Planchet], [18, Aubin, Chen & Dordan] and [17, *Tychastic Measure of Viability Risk. A Viabilist Portfolio Performance and Insurance Approach*]).

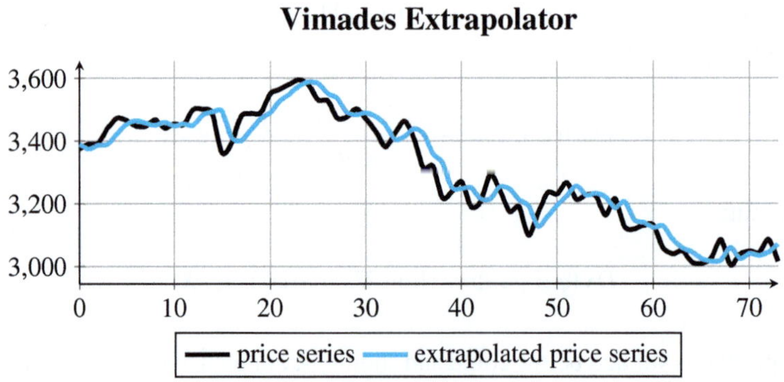

33 (VIMADES extrapolator) *The VIMADES extrapolator takes into account the velocity, the acceleration and the jerk of the history of an evolution for capturing its trend. Even though the extrapolation is good, it does not provide any explanation on the possible existence of patterns provided by differential equations. The figure displays an example of sliding extrapolation of the CAC 40 during the 2011 Crisis*[18–48]*, using the VIMADES extrapolator.*

Chapter 6
Endowing Fundamental Values: Willingness to Pay

6.1 A Static Viewpoint

The concept of *fundamental value* mentioned by *Turgot* in his letter[6–12] to *Hume* has known a recent revival under the name of "Willingness to Pay" for giving a value to public goods which cannot be exchanged, such as the evaluation of biodiversity (See [84, Hanemann], [78, Griffon], [125, Weber], [37, *Évaluation économique de la biodiversité*] de Brahic et J.-Ph. Terreaux and its bibliography, among an infinity of other publications on this topic).

"Willingness To Pay," respectively, "Willingness To Accept," is defined in the literature as *the maximum amount a person would be willing to pay (in units of numéraire) of an exchange of an "economic state" to receive (resp. accepting) the profit or avoid the sacrifice or something undesirable*, such as pollution.

One could, from a perspective that is not strictly "monetary," call this *fundamental value* "Willingness To Deprivation" (or "Willingness To Forego"), rather than "Willingness To Pay." In fact, the action of payment requires the introduction of a numéraire and of prices, which involve a "market mechanism" (Turgot's *exchange value*). Even though, if, in practice and in various empirical studies and surveys, for simplicity, the Willingness To Pay is measured in units of numéraire, it is not necessarily essential that the utility or cost functions involved are constructed from market mechanisms, of which we know very little of their operating modes, especially in a dynamical framework. Once obtained, this fundamental value should become a constraint in an exchange mechanism, in addition to scarcity constraints and those due to the desires of economic agents.

We should find a concept of time value (duration of deprivation) together with a concept of units of numéraire. Enquiries and investigations based on the questionnaires should involve also questions on *duration* of the processes, especially in environmental and biodiversity issues, where degradation is function of time (among other variables, not all economic) and where the duration could be also measured upstream by the repair mechanisms and downstream by the consequences of the deterioration of the environment. The size of a temporal window which

J.-P. Aubin, *Time and Money*, Lecture Notes in Economics and Mathematical Systems 670, DOI 10.1007/978-3-319-00005-3_6,

is computed is not given in advance, and offers an index of "sustainability" (defined as a capacity to "endure," having the same root as "duration"), among a long series of sustainability indicators. The indicator we use is derived from a potential patrimonial value in units of numéraire and provides *both* the duration (infinite for perennial durations) and the endowment to cover the transaction costs of the commodity.

A (static) economic perspective of *Willingness To Pay* involves as a datum a function $u : (x, w) \mapsto u(x, w)$ assessing or valuating each pair made of a "state" (or a commodity) $x \in \mathbb{R}^{\ell}$ and a parameter $w \in \mathbb{R}$ regarded as a "wealth" or as an income. If x_0 and w_0 denote the original state and its value, and x and w another state and its value, the question arises to compute the value w attributed to x in such a way that (x, w) *has the same utility than* (x_0, w_0). This means that w is a solution to the equation[1]

$$\mathbf{u}(x, w) = \mathbf{u}(x_0, w_0) \tag{6.1}$$

Then $\varpi(x; x_0, w_0) := w - w_0$ can be regarded as the *transaction cost* for obtaining x from x_0, the *"Willingness To Pay"* for exchanging x_0 with x defined implicitly as a solution to the equation

$$\mathbf{u}(x, w_0 + \varpi(x; x_0, w_0)) = \mathbf{u}(x_0, w_0) \tag{6.2}$$

In other words, the question arises to identifying two different pairs *having the same Willingness To Pay valuation*, and *compute the function* ϖ as a solution to this problem.

6.2 An Evolutionary Viewpoint

However, trades, transactions, etc. involve some underlying evolutionary (dynamical) process for exchanging an initial commodity x_0 with a new one, which requires the introduction of current time t ranging over a *temporal window* $[T - \Omega, T]$ of aperture Ω which should be computed.

Integrating from now on the income w as one of the components of economic state $x \in X := \mathbb{R}^{l}$, to say that one state $x(T)$ at the end of a temporal window $[T - \Omega, T]$ has the same "Willingness To Pay" value than the state $x(T - \Omega)$ at its

[1]Even when the set of solutions (x, w) contains more than one unique solution, inverse function theorems are available for providing the existence of solutions in a neighborhood of (x_0, w_0) and their stability (in the sense of pseudo-Lipschitz or Aubin property). See Chap. 3 of [24, *Set-Valued Analysis*] for a sufficient condition and Sect. 9.7 of [15, *Viability Theory. New Directions*] for a necessary and sufficient condition based on viability theory.

beginning after a succession of instantaneous transactions by an *evolutionary system* assumed to be known, encapsulate the same idea.

In the example above, (x, w) is regarded as the successor of (x_0, w_0). In this evolutionary context, this would mean that *the evolution between pairs* (x_0, w_0) *and* (x, w) *leaves constant the Willingness To Pay function* $u : (x, w) \mapsto u(x, w)$.

6.2.1 The Canonical Endowment of Fundamental (or WTP) Value

In the dynamical framework, instead of assuming that a *Willingness To Pay function* $u : (x, w) \mapsto u(x, w)$ is given as in the static case, we suggest to built it from the economic constraints, an evolutionary system governing the evolution of commodities and an (economic) potential function $U : x \in \mathbb{R}^\ell \mapsto U(x)$ assigning to each x its patrimonial value $U(x)$.

34 (Data of the Willingness to Pay Problem) *The data of the endowment problem we shall investigate are:*

1. *The* economic tube $t \in \mathbb{R} \rightsquigarrow K(t) \subset X$ *describing the time-dependent "economic environment"* $K(t) \subset X$ *of available and desired commodities;*
2. *The* departure set $\mathbb{D} \subset X$ *in which initial states are allowed to be chosen;*
3. *The state-dependent* return map $x \mapsto \mathbf{m}(x)$ *defining a* discount factor $\int_t^T \mathbf{m}(x(\tau))d\tau$ *from* t *up to time* T.

The return map is introduced for taking into account the "curse of the exponential": it should be chosen to 0 for replacing compounded interests by simple ones, or imposed as a constant. *The return function* $\mathbf{m}$ *is not an explanation, but a datum*, even though the actual rate $\mathbf{m}(x(t))$ depend on the yet unknown state $x(t)$. There is no economic reason at this stage why we could not take $\mathbf{m} \equiv 0$. However, there is no mathematical reason forbidding to study the general case.

We impose a finite bound $c < +\infty$ on the transactions $x'(t)$ of the commodities for describing the canonical dynamics of their evolution.

Definition 6.2.1 (Canonical Evolutionary System). These data being given, we associate with them the set $\mathscr{A}_c(\mathbb{D}, K(\cdot); T, x)$ of pairs $(\Omega, x(\cdot))$ made of durations $\Omega \geq 0$ and evolutions $x(\cdot)$ with velocity bounded by c, $0 \leq c < +\infty$, starting from a (nonempty) departure set $\mathbb{D} \subset X$ at time $T - \Omega$, viable in the time-dependent economic environment $t \mapsto K(t)$ and arriving at $x = x(T)$ at time T.

The behavioral assumption is described by the datum of the (economic) *potential function* $U : x \in X \mapsto U(x) \subset X$, describing the "patrimonial" or "intrinsic value" of a commodity. For simplicity of the exposition, we assume that U is continuously differentiable.

Definition 6.2.2 (Impetus of a Potential Value Function). The *impetus* $E(x(t))$ along an evolution $x(\cdot) : t \mapsto x(t)$ of the patrimonial value $U(x(t))$ is defined by

$$E(x(t)) \;:=\; \frac{d}{dt}U(x(t)) \;=\; \left\langle \frac{\partial U(x(t))}{\partial x}, x'(t) \right\rangle \tag{6.3}$$

where $x'(t)$ is regarded as the transaction of the commodity and where the gradient $p(t) := \dfrac{\partial U(x(t))}{\partial x}$ is regarded as a "price" (according to the marginal interpretation of prices).

Therefore, the impetus $E(x(t)) := \dfrac{d}{dt}U(x(t))$ can be interpreted as a *transaction potential*. The problem is to match the maximal impetus $\sup_{t\in[T-\Omega,T]} E(x(t))$ on the temporal window $[T-\Omega, T]$ of the endowment function $W : (T,x) \mapsto W(T,x)$ for which there exists at least one duration $\Omega \geq 0$ and one evolution $x(\cdot) \in \mathscr{A}_c(\mathbb{D}, K(\cdot); T, x)$ such that, for each $t \in [T-\Omega, T]$, $E(x)(t) \leq e^{-\int_t^T \mathbf{m}(x(\sigma))d\sigma} W(T,x)$. This leads to the following definition:

Definition 6.2.3 (The Endowment Function). The data[34–87] being given, the endowment function $W : (T,x) \mapsto W(T,x) \in \overline{\mathbb{R}} := \{-\infty\} \cup \mathbb{R} \cup \{+\infty\}$ is defined by

$$\begin{cases} W(T,x) \;:=\; \inf_{\{(\Omega,x(\cdot))\in\mathscr{A}_c(\mathbb{D},K(\cdot);T,x) \text{ such that } x(T-\Omega)\in\mathbb{D}\}} \\ \sup\limits_{t\in[T-\Omega,T]} \left(e^{\int_t^T \mathbf{m}(x(\sigma))d\sigma} \left\langle \dfrac{\partial U(x(t))}{\partial x}, x'(t) \right\rangle \right) \end{cases} \tag{6.4}$$

The endowment function W induces a "stratification" of the subset $K(T)$ into the subsets

$$\mathbf{K}(\sigma) := \{(T,x) \in \text{Graph}(K) \text{ such that } W(T,x) \leq \sigma\} \tag{6.5}$$

1. $\{(T,x)$ such that $-\infty < W(T,x) < 0\}$ is the subset of states $x \in K(T)$ which are *both economically* and *financially viable*.
2. $\{(T,x)$ such that $0 < W(T,x) < +\infty\}$ is the subset of states $x \in K(T)$ which are *economically viable* but *not financially viable*;
3. $\{(T,x)$ such that $W(T,x) := +\infty\}$ is the subset of states $x \in K(T)$ which are neither *economically* nor *financially viable*;

Warning 1. The *potential function* $U : x \mapsto U(x)$ is not the dynamical counterpart of the static utility function $u : (x, w) \mapsto u(x, w)$, the role of which is played by the *endowment function* $(T, x) \mapsto W(T, x)$. ■

6.2.2 The Viability Solution to the Endowment Problem

We shall state that the endowment function W coincides with the "viability solution" of the endowment problem that we have to define. For that purpose, we introduce the reverse *"characteristic control system"*

$$\begin{cases} (i) & \tau'(t) = -1 \\ (ii) & \overleftarrow{x}'(t) = -\upsilon(t) \\ (iii) & \overleftarrow{y}'(t) = -\mathbf{m}(\overleftarrow{x}(t))\overleftarrow{y}(t) \\ & \text{where the regulon } \overleftarrow{\upsilon}(t) \in \mathscr{U}(\overleftarrow{x}(t), \overleftarrow{y}(t)) \end{cases} \tag{6.6}$$

Let us introduce the extended functions

1. $\mathbf{k}$ defined by $\mathbf{k}_{-\infty}(t, x) = -\infty$ whenever $x \in K(t)$ and $\mathbf{k}_{-\infty}(t, x) = +\infty$ otherwise;
2. $\mathbf{c} := \mathbf{c}_{\mathbb{D}}$ where $\mathbf{c}_{\mathbb{D}}(t, x) := 0$ whenever $x \in \mathbb{D} \subset X$ (departure set) and $\mathbf{c}_{\mathbb{D}}(t, x) := +\infty$ otherwise.

The definition of a viability solution involves the concept of capture basin $\mathrm{Capt}_{(6.6)}(\mathscr{E}p(\mathbf{k}), \mathscr{E}p(\mathbf{c}))$ of a target, the epigraph $\mathscr{E}p(\mathbf{c})$ of the function $\mathbf{k}$ viable in the epigraph $\mathscr{E}p(\mathbf{c})$ of the function $\mathbf{c}$.

Definition 6.2.4 (The Viability Solution). The viability solution $V : (T, x) \mapsto V(T, x) \in \mathbb{R} \cup \{+\infty\}$ at time T and state x is defined by

$$V(T, x) := \inf_{(T,x,y) \in \mathrm{Capt}_{(6.6)}(\mathscr{E}p(\mathbf{k}), \mathscr{E}p(\mathbf{c}))} y \tag{6.7}$$

The principal theorem states that the endowment function is equal to the "viability solution" and inherits all the properties of capture basins:

Theorem 6.2.5 (Endowment Functions and Viability Solutions coincide). *The* viability solution V *defined by (6.7), p. 89, is equal to the* valuation function $W(\cdot, \cdot)$ *defined by (6.4), p. 88.*

We prove this statement in a slightly more general framework (Theorem 6.3.5, p. 99) below when we add other costs in the definition of the endowment function. The introduction of these costs may be useful in further studies.

The solution to this problem is given *in terms of subsets*: the capture basin $\mathrm{Capt}(K, C)$, which is the epigraph of the endowment function, and the graph of the retroaction map (graphical approach of maps). Instead of handling functions as in classical analysis, *viability theory manipulates subsets as in set-valued analysis* (see [24, Aubin & Frankowska] or [114, Rockafellar & Wets] for instance), and, in particular, graphs of maps and epigraphs of real-valued functions (see also [11, *Mutational and morphological analysis: tools for shape regulation and morphogenesis*]).

These maps are no longer obtained through analytical formulas, but can be computed in the framework of "set-valued numerical analysis." The viability algorithms and software handle at each iteration subsets instead of vectors. They are subject to the "dimensionality curse," which limits the dimension of the problem according to the capacities of the hardware.

6.2.3 Optimality Theorems on Endowment Functions

We state the main result which will be derived from the more general Theorems 6.3.2, p. 98 and 6.3.3, p. 98[2]:

Theorem 6.2.6 (Optimal Evolutions). *There exists at least one optimal evolution $(\Omega_{\mathrm{opt}}, x_{\mathrm{opt}}(\cdot)) \in \mathscr{A}_c(\mathbb{D}, K(\cdot); T, x)$ arriving at x at time T. The following conditions are equivalent:*

1. *The evolution $(\Omega_{\mathrm{opt}}, x_{\mathrm{opt}}(\cdot)) \in \mathscr{A}_c(\mathbb{D}, K(\cdot); T, x)$ is optimal in the sense that*

$$W(T,x) \;=\; \sup_{t \in [T-\Omega_{\mathrm{opt}}, T]} \left(e^{\int_t^T \mathbf{m}(x_{\mathrm{opt}}(\sigma))d\sigma} \left\langle \frac{\partial U(x(t))}{\partial x}, x'_{\mathrm{opt}}(t) \right\rangle \right) \tag{6.8}$$

2. *The evolution $(\Omega_{\mathrm{opt}}, x_{\mathrm{opt}}(\cdot)) \in \mathscr{A}_c(\mathbb{D}, K(\cdot); T, x)$ satisfies the* dynamic optimality *property along the optimal evolution: whenever $W(T-\Omega, x(T-\Omega)) > -\infty$,*

$$\forall\, t \in [T-\Omega_{\mathrm{opt}}, T],\; e^{-\int_t^T \mathbf{m}(x_{\mathrm{opt}}(\sigma))d\sigma} W(t, x_{\mathrm{opt}}(t)) \;=\; W(T,x) \tag{6.9}$$

This means that the discounted value of the endowment remains constant along optimal evolutions.

[2]This is, in essence, another formulation of the Isaacs-Bellman *dynamic optimality property* for control problems, known since at least *Constantin Carathéodory* in calculus of variations (see Theorem 6.3.3, p. 98).

Furthermore, optimal evolutions are solution to the differential equation

$$\forall\, t \in [T - \Omega_{\text{opt}}, T],\ \ \frac{d}{dt} W(t, x_{\text{opt}}(t)) \ = \ \mathbf{m}(x(t)) W(t, x_{\text{opt}}(t)) \qquad (6.10)$$

Remark (Viability and Optimality). Viable evolutions $x(\cdot)$ are the ones such that $(t, x(t), y(t))$ are viable in the epigraph $\mathscr{E}p(W)$ of the endowment function W whereas optimal evolutions $x_{\text{opt}}(\cdot)$ are the ones such that $(t, x_{\text{opt}}(t), y_{x_{\text{opt}}(\cdot)}(t))$ are viable in the graph Graph(W).

This means that there may exist viable (but non-optimal) evolutions $t \mapsto (t, x(t), y_{x(\cdot)}(t))$ satisfying $y_{x(\cdot)}(t) > W(t, x(t))$ on some interval. They are viable in the epigraph of the endowment function W, but not in its graph. Therefore, *the evolution $x(\cdot)$ is viable but not optimal.* Observe that we can take any velocity $x(\cdot)$ viable in the tube $t \rightsquigarrow K(t)$ since the sign of $y'_{x(\cdot)}(t)$ is arbitrary as long as $y_{x(\cdot)}(t) > W(t, x(t))$. However, as soon as for some $t^\star \le T$, the endowment $y_{x(\cdot)}(t^\star) = W(t^\star, x(t^\star))$ reaches the graph of W, *the viable evolution $x(\cdot)$ becomes optimal, $x(\cdot) := x_{\text{opt}}(\cdot)$, satisfies $x_{\text{opt}}(t) = W(t, x_{\text{opt}}(t))$ until it arrives at* $(T, x(T), W(T, x(T)))$. They are obliged to be governed by "optimal velocities" provided by the regulation map we are about to construct from the endowment function.[3]

In other words, the function $t \mapsto W(t, x_{\text{opt}}(t), y_{x_{\text{opt}}(\cdot)}(t))$ can be regarded as a *optimality threshold.* ■

Remark. The required property (6.1), p. 86, in the static case (and thus, with no interest rate) is that the endowment function remains constant along the evolution of optimal evolutions:

$$\forall\, t \in [T - \Omega_{\text{opt}}, T],\ \ W(t, x_{\text{opt}}(t)) \ := \ W(T, x)$$

thanks to the dynamic optimality property. Hence *the endowment function is a reasonable candidate for playing the role of a Willingness To Pay function* as it is defined in the literature.

When the economic state is a pair $(x, w) \in X \times \mathbb{R}_+$ where $x \in X$ is a commodity and w a wealth, for any $t \in [T - \Omega_{\text{opt}}, T]$, $x_{\text{opt}}(T) - x_{\text{opt}}(t) = \int_t^T x'_{\text{opt}}(\tau) d\tau$ is the *transaction* between t and T and

$$\forall\, t \in [T - \Omega_{\text{opt}}, T],\ \ \varpi(x; x_{\text{opt}}(t), w_{\text{opt}}(t)) \ := \ \int_t^T w'_{\text{opt}}(\tau) d\tau \qquad (6.11)$$

[3]This is the consequence of the Quincampoix Theorem on the semi-permeability barrier property of viability kernels.

is its *transaction cost*, regarded as the Willingness To Pay in the static case which motivated this study. ■

Remark. The use of the supremum $\sup_{t\in[T-\Omega,T]}$ in (6.4), p. 88, is not rigourously mathematically correct, since the velocities $x'(\cdot)$ are not necessarily continuous functions, but only measurable and bounded. The correct notation should be $\sup \text{ess}_{t\in[T-\Omega,T]}$ (essential supremum). If we want to obtain a continuously differentiable evolution $x(\cdot)$, we may require that the accelerations are also bounded. In this case, we can define the amount $W(T,x,v)$ where we require that at terminal time T, conditions $x(T) = x$ *and* $x'(T) = v$ are prescribed. At this stage, this technical intricacy (see [13, Aubin] and Chap. 6 of [15, *Viability Theory, New Directions*]) is bypassed for the sake of simplicity of the exposition since this sin against mathematics does not interfere with the economic interpretation. ■

6.2.4 Microscopic and Macroscopic Properties of the Endowment Function

We state two complementary mathematical properties of the endowment function, interesting for their interpretation as mathematical metaphors, but not practically useful since *the viability algorithms provide both the endowment function and the regulation map governing the evolution of optimal evolutions* $x(\cdot) := x_{\text{opt}}(\cdot)$ *arriving at* $x(T)$ *on the interval* $[T-\Omega_{\text{opt}},T]$.

In order to take into account the particular shape of the endowment function involving the impetuses $\left\langle \frac{\partial U(x(t))}{\partial x}, x'(t) \right\rangle$ on the interval $[T-\Omega,T]$, we introduce the following set-valued map $\mathscr{U}$ defined by

$$\mathscr{U}(x,y) := \left\{ v \in cB \text{ such that } \left\langle \frac{\partial U(x)}{\partial x}, v \right\rangle \leq y \right\} \tag{6.12}$$

providing the bounded velocities such that their transactions costs are bounded by y.

The knowledge of the endowment function allows us to introduce

1. the *"Hamiltonian" function* $\mathbf{I}_c^\star$ defined by

$$\mathbf{I}_c^\star\left(\frac{\partial U(x)}{\partial x}; \frac{\partial W(t,x)}{\partial x}, W(t,x)\right) := \inf_{v\in\mathscr{U}(x,W(t,x))} \left\langle \frac{\partial W(t,x)}{\partial x}, v \right\rangle \tag{6.13}$$

2. the (possibly empty) set-valued *regulation map* R_c defined by

$$\begin{cases} R_c\left(\frac{\partial U(x)}{\partial x}; \frac{\partial W(t,x)}{\partial x}, W(t,x)\right) := \{v \in \mathscr{U}(x, W(T,x)) \text{ such that} \\ \left\langle \frac{\partial W(t,x)}{\partial x}, v \right\rangle = \mathbf{I}_c^\star\left(\frac{\partial U(x)}{\partial x}; \frac{\partial W(t,x)}{\partial x}, W(t,x)\right)\right\} \end{cases} \tag{6.14}$$

The Hamiltonian function $\mathbf{l}_c^\star$ and the regulation map R_c depend on the endowment function $W(t,x)$, the price $\frac{\partial W(t,x)}{\partial x}$ derived from the endowment function W, and the given (or advised) price $\frac{\partial U(x)}{\partial x}$ obtained from the datum of the potential function U.

Knowing the endowment function $(t,x) \mapsto W(t,x)$ and assuming that it is continuously differentiable, we obtain

1. The *Microscopic Property*: optimal evolutions $x(\cdot) := x_{\text{opt}}(\cdot)$ are governed by the differential inclusion

$$\forall\, t \in [T-\Omega, T], \;\; x'(t) \;\in\; R_c\left(\frac{\partial U(x(t))}{\partial x}; \frac{\partial W(t,x(t))}{\partial x}, W(t,x(t))\right) \tag{6.15}$$

   satisfying the terminal condition $x(T) = x$.
2. The *Macroscopic Property*: The endowment function $W(t,x)$ is a solution to the Hamilton–Jacobi equation

$$\frac{\partial W(t,x)}{\partial t} \;=\; \mathbf{l}_c^\star\left(\frac{\partial U(x)}{\partial x}; \frac{\partial W(t,x)}{\partial x}, W(t,x)\right) + \mathbf{m}(x)W(t,x) \tag{6.16}$$

   satisfying the *boundary condition* $W(t,x(t)) = 0$ whenever $x(t) \in \mathbb{D}$ belongs to departure set.

These two seducing mathematical properties are consequences of the Viability and Invariance Theorems, as the proof of the existence of the Walras equilibrium is the consequence of the Brouwer Fixed Point Theorem. In both cases, they are technically involved. The proofs are not duplicated because microscopic and macroscopic properties are only examples of Theorems 17.4.1, p. 699, and 17.4.2, p. 701, of [15, *Viability Theory, New Directions*]. We mention also that we do not need to assume that the endowment function is continuously differentiable, but only lower semicontinuous and then replace the partial derivatives by generalized gradients.[4]

However, the idea behind these technicalities are quite simple, and we explain them formally.

The Formal Story. Starting from property (6.10), p. 91

$$\forall\, t \in [T-\Omega_{\text{opt}}, T], \;\; \frac{d}{dt}W(t, x_{\text{opt}}(t)) \;=\; \mathbf{m}(x(t))(W(t, x_{\text{opt}}(t)))$$

[4]When the endowment function is only lower semicontinuous, the viability solution coincides with the Barron-Jensen/Frankoska viscosity solutions (see [28, Barron, Jensen] and [66–69, Frankowska]). We do not elaborate these generalizations, since economic interpretation of this Hamilton–Jacobi equation do not play a major role in this study.

we infer that, along an optimal evolution $x(t) := x_{\text{opt}}(t)$

$$\frac{\partial W(t,x(t))}{\partial t} + \left\langle \frac{\partial W(t,x(t))}{\partial x}, x'(t) \right\rangle = \mathbf{m}(x(t))(W(t,x(t)))$$

We observe that

$$\left\langle \frac{\partial W(t,x(t))}{\partial x}, x'(t) \right\rangle = \mathbf{l}_c^\star \left(\frac{\partial U(x)}{\partial x}; \frac{\partial W(t,x)}{\partial x}, W(t,x) \right) \tag{6.17}$$

from which we infer that

$$x'(t) \in R_c \left(\frac{\partial U(x)}{\partial x}; \frac{\partial W(t,x)}{\partial x}, W(t,x) \right) \tag{6.18}$$

and

$$\begin{cases} \dfrac{\partial W(t,x(t))}{\partial t} - \mathbf{m}(x(t))W(t,x(t)) = \\ \left\langle \dfrac{\partial W(t,x(t))}{\partial x}, x'(t) \right\rangle = \mathbf{l}_c^\star \left(\dfrac{\partial U(x(t))}{\partial x}; \dfrac{\partial W(t,x(t))}{\partial x}, W(t,x(t)) \right) \end{cases} \tag{6.19}$$

which justify both *the microscopic and macroscopic properties*. The proof is a slight modification of the proofs of Theorems 13.6.2, p. 542 and 13.10.2, p. 558, of [15, *Viability Theory. New Directions*]. ■

Remark. Setting

$$\Pi := \frac{\partial U(x)}{\partial x} \text{ and } \Delta := \frac{\partial W(t,x)}{\partial x}$$

we observe that $\dfrac{\Pi}{\|\Pi\|^2}$ and $\Delta - \dfrac{\langle \Pi, \Delta \rangle}{\|\Pi\|^2}\Pi$ are orthogonal since

$$\left\langle \frac{\Pi}{\|\Pi\|^2}, \Delta - \frac{\langle \Pi, \Delta \rangle}{\|\Pi\|^2}\Pi \right\rangle = 0 \tag{6.20}$$

We define the regulation maps

$$R_\mu(\Pi, \Delta; y) := \left(\frac{y - \mu \langle \Pi, \Delta \rangle}{\|\Pi\|^2} \right) \Pi + \mu\Delta = y\frac{\Pi}{\|\Pi\|^2} + \mu \left(\Delta - \frac{\langle \Pi, \Delta \rangle}{\|\Pi\|^2}\Pi \right) \tag{6.21}$$

The term $\Delta - \frac{\langle \Pi, \Delta \rangle}{\|\Pi\|^2} \Pi$ can be interpreted as a correction mechanism (defining a tâtonnement process) between the price $\Pi := \frac{\partial U(x)}{\partial x}$ advised by the potential function and the price $\Delta := \frac{\partial W(t,x)}{\partial x}$ corrected by the endowment function.

We observe that for every μ,

$$\big\langle \Pi, R_\mu(\Pi, \Delta; y)\big\rangle = y \tag{6.22}$$

Let us choose $\mu_c(\Pi, \Delta; y)$ to be the solution to

$$\big\| R_{\mu_c(\Pi,\Delta;y)}(\Pi, \Delta; y)\big\| = c \tag{6.23}$$

Hence

$$R_c(\Pi, \Delta; y) = R_{\mu_c(\Pi,\Delta;y)}(\Pi, \Delta; y)$$

minimizes the linear form $v \mapsto \langle \Delta, v \rangle$ over the subset $\mathscr{U}_\Pi(x, y) = \{v \in cB$ and $\langle \Pi, v \rangle \leq y\}$. ■

6.2.5 *The Endowment Effect*

For the sake of simplicity, we assumed that the endowment function is continuously differentiable. Hence the *Willingness To Pay* value of the transaction v defined by $\left\langle \frac{\partial W(t,x)}{\partial x}, v \right\rangle$ is equal to the *Willingness To Accept* value $-\left\langle \frac{\partial W(t,x)}{\partial x}, -v \right\rangle$ of the transaction $-v$. This is a consequence of the linearity of the derivative of a continuous differentiable function.

However, when the endowment function W is only lower semicontinuous, this equality is no longer necessarily true and may provide a justification of the "endowment effect" we are about to define.

Although we avoided, at this stage, the burden of replacing the derivative $DW(t,x)(v) := \left\langle \frac{\partial W(t,x)}{\partial x}, v \right\rangle$ for denoting the gradient of $W(t,x)$ at x in the direction v by the (convexified contingent) epiderivative $u \mapsto D_\uparrow W(t,x)(v)$, which is no longer linear, but convex and lower semicontinuous (see Footnote 6, p. 99), we need it shortly. We define in this case the endowment function by

$$\begin{cases} W(T,x) := \inf_{\{(\Omega,x(\cdot))\in\mathscr{A}_c(\mathbb{D},K(\cdot);T,x) \text{ such that } x(T-\Omega)\in\mathbb{D}\}} \\ \sup_{t\in[T-\Omega,T]} e^{\int_t^T \mathbf{m}(x(\sigma))d\sigma} D_\uparrow W(t,x(t))(1,x'(t)) \end{cases} \tag{6.24}$$

When W is lower semicontinuous and sleek, we obtain only the inequality[5]

$$D_{\uparrow}(V)(x)(v) \leq -D_{\uparrow}V(x)(-v) \tag{6.25}$$

Therefore, the *Willingness To Pay* value $D_{\uparrow}(V)(x)(v)$ of the transaction v is smaller than or equal to the *Willingness To Accept* value $-D_{\uparrow}V(x)(-v)$ of the transaction $-v$.

This may provide an explanation to the endowment effect:

35 (Endowment Effect) *The specialists of this topic postulate that agents evaluate asymmetrically the values of the transactions: the Willingness To Pay value of any transaction should be lower than the Willingness To Accept it, instead of being equal as standard economics require. So, the endowment effect appears whenever the endowment function is not differentiable at a given state $x(t)$ (for which the willingness to pay and to accept are different), but only lower semicontinuous and sleek.*

This property has been coined the *endowment effect* by *Richard Thaler* in 1980. Psychologists observed the difference between consumers WTP and WTA not only for *Homo oeconomicus*, but for children (see [85, Harbaugh, Krause & Vesterlund]) and ... apes (see [92, Kanngiesser, Santos, Hood & Call] and [38, Brosnan, Jones, Gardner, Lambeth & Schapiro]).

6.3 General Valuation Functions

6.3.1 Adding Other Costs to the Endowment

We introduce a new functional $A : x(\cdot) \mapsto A(x(\cdot))$ on evolutions $x(\cdot)$ defined by

$$A(x(\cdot))(t) := \alpha e^{\int_t^T \mathbf{m}(x(\cdot))(\sigma)d\sigma} U(x(t)) + \int_t^T e^{\int_\tau^T \mathbf{m}(x(\cdot))(\sigma)d\sigma} \mathbf{l}(x'(\tau))d\tau - \alpha U(x) \tag{6.26}$$

involving the patrimonial value $U : x \mapsto U(x)$ (when $\alpha \neq 0$) and the cumulation of discounted transition costs $\mathbf{l}(x'(t))$.

[5] See Sect. 6.1, p. 219 of [24, *Set-Valued Analysis*]. Sleekness plays the role of continuous differentiability.

Hence $y(\cdot) := y_{x(\cdot)}(\cdot)$ defined by

$$\begin{cases} y_{x(\cdot)}(t) & = e^{-\int_t^T \mathbf{m}(x(\sigma))d\sigma} y - A(x(\cdot))(t) \\ = e^{-\int_t^T \mathbf{m}(x(\sigma))d\sigma} [y - \alpha U(x(t))] + \alpha U(x) - \int_t^T e^{\int_\tau^T \mathbf{m}(x(\sigma))d\sigma} \mathbf{l}(x'(\tau))d\tau \end{cases} \tag{6.27}$$

is the solution to differential equation

$$y'(t) = \mathbf{m}(x(t))y(t) + \alpha U(x(t)) + \alpha \left\langle \frac{\partial U(x(t))}{\partial x}, x'(t) \right\rangle + \mathbf{l}(x'(t)) \tag{6.28}$$

arriving at y at time T.

For simplifying the notations, we no longer mention explicitly the economic tube nor the departure set $\mathbb{D}$. They appear in the definitions of the transaction cost function $\mathbf{k}$ and objective function $\mathbf{c}$ we are about to introduce. Then $\mathscr{A}_c(T,x)$ *denotes the set of evolutions with bounded velocities (by* c*) arriving at* x *at time* T.

Definition 6.3.1 (The Valuation Function). The valuation function is defined by

$$\begin{cases} W(T,x) := \inf_{x(\cdot)\in\mathscr{A}_c(T,x)} \inf_{\Omega\in[0,T]} \\ \max\left[e^{\int_{T-\Omega}^T \mathbf{m}(x(\sigma))d\sigma} \mathbf{c}(T-\Omega, x(T-\Omega)) + A(x(\cdot))(T-\Omega), \right. \\ \left. \sup_{t\in[T-\Omega,T]} e^{\int_t^T \mathbf{m}(x(\sigma))d\sigma} \max\left(\mathbf{k}(t,x(t)), \frac{d}{dt}U(x(t))\right) + A(x(\cdot))(t)\right] \end{cases} \tag{6.29}$$

The endowment function $W(\cdot)$ involves the maximum of the instantaneous functional

$$\mathbf{J}_\mathbf{c}(T) := e^{\int_{T-\Omega}^T \mathbf{m}(x(\sigma))d\sigma} \mathbf{c}(T-\Omega, x(T-\Omega)) + A(x(\cdot))(T-\Omega) \tag{6.30}$$

depending on the objective cost function $\mathbf{c}(\cdot)$ and of the constraint cost functional

$$\sup_{t\in[T-\Omega,T]} e^{\int_t^T \mathbf{m}(x(\sigma))d\sigma} \max\left(\mathbf{k}(t,x(t)), \frac{d}{dt}U(x(t))\right) + A(x(\cdot))(t) \tag{6.31}$$

depending on the constraint cost function $\mathbf{k}(\cdot)$.

Remark. The endowment function is a particular case of the valuation function of Definition 6.3.1, p. 97, when we choose $\alpha = 0$, $\mathbf{l}(\cdot) \equiv 0$, the function $\mathbf{k} := \mathbf{k}_{-\infty}$ such that $\mathbf{k}_{-\infty}(t,x) = -\infty$ whenever $x \in K(t)$ and $\mathbf{k}_{-\infty}(t,x) = +\infty$ otherwise and $\mathbf{c} := \mathbf{c}_\mathbb{D}$ where $\mathbf{c}_\mathbb{D}(t,x) := 0$ whenever $x \in \mathbb{D} \subset X$ (departure set) and $\mathbf{c}_\mathbb{D}(t,x) := +\infty$ otherwise. ∎

Theorem 6.3.2 (Viable and Optimal Evolutions Coïncide). *When $(T, x, W(T, x))$ belongs to the epigraph of the valuation function $(T, x) \mapsto W(T, x)$ (which is the case if the epigraph is closed), then there exist at least one optimal duration Ω_{opt} and one evolution $x_{\text{opt}}(\cdot) \in \mathscr{A}_c(T, x)$ defined on the temporal window $[T - \Omega_{\text{opt}}, T]$ and arriving at time x at time T. The following conditions are equivalent:*

1. *The evolution $x_{\text{opt}}(\cdot) \in \mathscr{A}_c(T, x)$ is* optimal*:*

$$\left\{ \begin{array}{l} W(T, x) \\ = \max \left[e^{\int_{T-\Omega}^{T} \mathbf{m}(x_{\text{opt}}(\sigma))d\sigma} \mathbf{c}(T - \Omega, x_{\text{opt}}(T - \Omega_{\text{opt}})) \right. \\ +A(x_{\text{opt}}(\cdot))(T - \Omega_{\text{opt}}), \\ \sup_{t \in [T-\Omega_{\text{opt}}, T]} \left(e^{\int_t^T \mathbf{m}(x(\sigma))d\sigma} \max \left(\mathbf{k}(t, x(t)), \frac{d}{dt} U(x_{\text{opt}}(t)) \right) \right. \\ \left. +A(x_{\text{opt}}(\cdot))(t) \right) \Big] \end{array} \right. \quad (6.32)$$

2. *The evolution $x_{\text{opt}}(\cdot)$ is* viable*, in the sense that $t \mapsto (t, x_{\text{opt}}(t), y_{x_{\text{opt}}(\cdot)}(t))$ arriving at $(T, x, W(T, x))$ at time T is viable in $\mathscr{E}p(W)$ on $[T - \Omega_{\text{opt}}, T]$.*

The optimal solution satisfies the dynamic optimality property:

Theorem 6.3.3 (Dynamic Optimality Property of the Valuation Function). *Let us consider a viable evolution $x(\cdot)$ with which we associate the duration $\Omega_{\mathbf{k}} \in [0, \Omega_{\text{opt}}]$ when $W(T - \Omega_{\mathbf{k}}, x(T - \Omega_{\mathbf{k}})) = \mathbf{k}(T - \Omega_{\mathbf{k}}, x(T - \Omega_{\mathbf{k}}))$. Therefore, the valuation function satisfies the* dynamic optimality *along the optimal evolution:*

$$\forall\, t \in [T - \Omega_{\mathbf{k}}, T], \;\; e^{-\int_t^T \mathbf{m}(x(\sigma))d\sigma} W(t, x(t)) + A(x(\cdot))(t) \;=\; W(T, x) \quad (6.33)$$

Furthermore, it is the solution to the differential equation: $\forall\, t \in [T - \Omega_{\mathbf{k}}, T]$,

$$\left\{ \begin{array}{l} \frac{d}{dt} W(t, x(t)) \;=\; \mathbf{m}(x(t))\,(W(t, x(t)) - \alpha U(x(t))) \\ +\mathbf{l}(x'(t)) + \alpha \left\langle \frac{\partial U(x(t))}{\partial x}, x'(t) \right\rangle \end{array} \right. \quad (6.34)$$

6.3.2 *The Viability Solution*

Recall that[6] we set

$$\mathscr{U}(x,y) := \left\{ v \in cB \text{ and } \left\langle \frac{\partial U(x)}{\partial x}, v \right\rangle \le y \right\} \tag{6.36}$$

which is a Marchaud set-valued map whenever the potential function U is continuously differentiable (or lower semicontinuous and sleek). Let us consider the reverse *"characteristic control system"*

$$\begin{cases} (i) \ \tau'(t) = -1 \\ (ii) \ \overleftarrow{x}'(t) = -v(t) \\ (iii) \ \overleftarrow{y}'(t) = -\mathbf{m}(\overleftarrow{x}(t))\left(\overleftarrow{y}(t) + \alpha U(\overleftarrow{x}(t))\right) \\ \quad -\mathbf{l}(v(t)) + \alpha \left\langle \frac{\partial U(\overleftarrow{x}(t))}{\partial x}, v(t) \right\rangle \\ \quad \text{where the regulon } v(t) \in \mathscr{U}(x(t), y(t)) \end{cases} \tag{6.37}$$

Definition 6.3.4 (The Viability Solution). The viability solution $V(T,x) \in \mathbb{R} \cup \{+\infty\}$ at time T and state x is defined by

$$V(T,x) := \inf_{(T,x,y) \in \mathrm{Capt}_{(6.37)}(\mathscr{E}p(\mathbf{k}), \mathscr{E}p(\mathbf{c}))} y \tag{6.38}$$

Theorem 6.3.5 (The Viability Solution). *The* viability solution V *defined by (6.38), p. 99, is equal to the* valuation function $W(\cdot,\cdot)$ *defined by (6.29), p. 97.*

Proof. We adapt the proof of Theorem 3.2.2, p. 51 to this nondeterministic case.

1. For proving the theorem, we perform the change of the variable $s = T - t$. We observe that

$$\int_0^t \mathbf{m}(\overleftarrow{x}(\tau))d\tau = \int_{T-t}^T \mathbf{m}(x(\sigma))d\sigma \tag{6.39}$$

---

[6]If the potential function U is only lower semicontinuous and sleek, we set

$$\mathscr{U}(x,y) := \{v \in cB \text{ and } D_{\uparrow}U(x)(v) \le y\} \tag{6.35}$$

Setting

$$A(\overleftarrow{x}(\cdot))(t) := \alpha e^{\int_0^t \mathbf{m}(\overleftarrow{x}(\sigma))d\sigma} U(\overleftarrow{x}(t)) + \int_0^t e^{\int_0^\tau \mathbf{m}(\overleftarrow{x}(\sigma))d\sigma} \mathbf{l}(-\overleftarrow{x}'(\tau))d\tau - \alpha U(x) \tag{6.40}$$

the function $\overleftarrow{y}(\cdot)$ defined by

$$\begin{cases} \overleftarrow{y}(t) := \overleftarrow{y}_{(\overleftarrow{x}(\cdot),y)}(t) := e^{-\int_0^t \mathbf{m}(\overleftarrow{x}(\sigma))d\sigma} \left[y - A(\overleftarrow{x}(\cdot))(t) \right] \\ = e^{-\int_0^t \mathbf{m}(\overleftarrow{x}(\sigma))d\sigma} \Big(y + \alpha U(x) - \alpha e^{\int_0^t \mathbf{m}(\overleftarrow{x}(\sigma))d\sigma} U(\overleftarrow{x}(t)) \\ -e^{\int_0^\tau \mathbf{m}(\overleftarrow{x}(\sigma))d\sigma} \mathbf{l}(-\overleftarrow{x}'(\tau))d\tau \Big) \end{cases} \tag{6.41}$$

is the solution to the differential equation

$$\overleftarrow{y}'(t) = -\mathbf{m}(\overleftarrow{x}(t)) \left(\overleftarrow{y}(t) + \alpha U(\overleftarrow{x}(t)) \right) - \mathbf{l}(-\overleftarrow{x}'(t)) - \alpha \left\langle \frac{\partial U(\overleftarrow{x}(t))}{\partial x}, \overleftarrow{x}'(t) \right\rangle \tag{6.42}$$

starting at 0 (since $U(\overleftarrow{x}(0)) = U(x)$).

Hence the formula defining $W(T,x)$ in the form (6.29), p. 97 can be rewritten as

$$\begin{cases} W(T,x) := \inf_{\overleftarrow{x}(\cdot)\in\mathcal{S}_c(x)} \inf_{\Omega\in[0,T]} \max \Big[\\ e^{\int_0^{\varpi} \mathbf{m}(\overleftarrow{x}(\sigma))d\sigma} \mathbf{c}(T-\Omega, \overleftarrow{x}(\Omega)) + A(\overleftarrow{x}(\cdot))(\Omega), \\ \sup_{t\in[0,\Omega]} e^{\int_0^{\Omega} \mathbf{m}(x(\sigma))d\sigma} \max \left(\mathbf{k}(T-t, \overleftarrow{x}(t)), \frac{d}{dt} U(\overleftarrow{x}(t)) \right) + A(\overleftarrow{x}(\cdot))(t) \Big] \end{cases} \tag{6.43}$$

2. Let (T,x,y) belong to the capture basin $\mathrm{Capt}_{(6.37)}(\mathcal{E}p(\mathbf{k}), \mathcal{E}p(\mathbf{c}))$. This means that there exist evolutions $t \mapsto T-t$, $t \mapsto \overleftarrow{x}(t)$ and $t \mapsto \overleftarrow{y}(t) := \overleftarrow{y}_{(\overleftarrow{x}(\cdot),y)}(t)$ defined by (6.41), p. 100 starting at (T,x,y) and a time $\Omega^{\flat} \geq 0$ which is viable in $\mathcal{E}p(\mathbf{k})$ until it reaches $\mathcal{E}p(\mathbf{c})$ at time $\Omega^{\flat}$.

   This implies that

   a. $(\overleftarrow{\tau}(\Omega^{\flat}), \overleftarrow{x}(\Omega^{\flat}), \overleftarrow{y}(\Omega^{\flat}))$ belongs to the epigraph of $\mathbf{c}$ and

$$\mathbf{c}(T-\Omega^{\flat}, \overleftarrow{x}(\Omega^{\flat})) \leq \overleftarrow{y}(\Omega^{\flat}) = e^{-\int_0^{\Omega^{\flat}} \mathbf{m}(x(\sigma))d\sigma} \left(y - A(\overleftarrow{x}(\cdot))(t) \right) \tag{6.44}$$

   i.e., that

$$e^{\int_0^{\Omega^{\flat}} \mathbf{m}(x(\sigma))d\sigma} \mathbf{c}(T-\Omega^{\flat}, \overleftarrow{x}(\Omega^{\flat})) + A(\overleftarrow{x}(\cdot))(\Omega^{\flat}) \leq y \tag{6.45}$$

b. The solution is viable in the epigraph of the reverse constraint transaction cost on the interval $[0, \Omega^\flat]$: $\forall\, t \in [0, \Omega^\flat]$,

$$\mathbf{k}(T-t, \overleftarrow{x}(t)) \le \overleftarrow{y}(t) = e^{-\int_0^t \mathbf{m}(\overleftarrow{x}(\sigma))d\sigma}\left(y - A(\overleftarrow{x}(\cdot))(t)\right) \tag{6.46}$$

i.e.,

$$\sup_{t\in[0,\Omega^\flat]}\left(e^{\int_0^t \mathbf{m}(x(\sigma))d\sigma}\mathbf{k}(T-t, \overleftarrow{x}(t)) + A(\overleftarrow{x}(\cdot))(t)\right) \le y \tag{6.47}$$

c. Furthermore, the definition of the characteristic control system (6.37), p. 99, implies that the velocity $\overleftarrow{x}'(t) = -v(t)$ has a norm bounded by c and satisfies $-\left\langle \frac{\partial U(\overleftarrow{x}(t))}{\partial x}, \overleftarrow{x}'(t)\right\rangle \le \overleftarrow{y}(t)$. Therefore

$$\sup_{t\in[0,\Omega^\flat]}\left(-e^{\int_0^t \mathbf{m}(x(\sigma))d\sigma}\left\langle \frac{\partial U(\overleftarrow{x}(t))}{\partial x}, \overleftarrow{x}'(t)\right\rangle + A(\overleftarrow{x}(\cdot))(t)\right) \le y \tag{6.48}$$

Hence, putting together these three inequalities, we obtain

$$\begin{cases} y \ge W(T,x) = \max\left[e^{\int_0^{\Omega^\flat} \mathbf{m}(x(\sigma))d\sigma}\mathbf{c}(T-\Omega^\flat, \overleftarrow{x}(\Omega^\flat)) + A(\overleftarrow{x}(\cdot))(\Omega^\flat),\right. \\ \sup_{t\in[0,\Omega^\flat]}\left(e^{\int_0^t \mathbf{m}(x(\sigma))d\sigma}\max\left[\mathbf{k}(T-t, \overleftarrow{x}(t)), -\left\langle \frac{\partial U(\overleftarrow{x}(t))}{\partial x}, \overleftarrow{x}'(t)\right\rangle\right]\right. \\ \left. +A(\overleftarrow{x}(\cdot))(t)\right) \end{cases} \tag{6.49}$$

3. Taking the infimum over the subset of y such that $(T,x,y) \in \mathrm{Capt}_{(6.37)}(\mathscr{E}p(\mathbf{k}), \mathscr{E}p(\mathbf{c}))$, we infer that the viability solution $W(T,x) \le V(T,x)$.
4. Conversely, let us take any $\varepsilon > 0$ and choose an evolution $\overleftarrow{x}_\varepsilon(\cdot) \in \mathscr{S}_c(x)$ and $\Omega^\flat_\varepsilon$ such that, by the very definition of the infimum,

$$\begin{cases} W(T,x) + \varepsilon \ge \max\left[e^{\int_0^{\Omega^\flat_\varepsilon} \mathbf{m}(x_\varepsilon(\sigma))d\sigma}\mathbf{c}(T-\Omega^\flat_\varepsilon, x(\Omega^\flat_\varepsilon))A(\overleftarrow{x}_\varepsilon(\cdot))(\Omega^\flat_\varepsilon),\right. \\ \sup_{t\in[0,\Omega^\flat_\varepsilon]}\left(e^{\int_0^t \mathbf{m}(x_\varepsilon(\sigma))d\sigma}\max\left[\mathbf{k}(T-t, \overleftarrow{x}_\varepsilon(t)), -\left\langle \frac{\partial U(\overleftarrow{x}_\varepsilon(t))}{\partial x}, \overleftarrow{x}'_\varepsilon(t)\right\rangle\right]\right. \\ \left. +A(\overleftarrow{x}_\varepsilon(\cdot))(t)\right) \end{cases} \tag{6.50}$$

Introducing the function

$$\overleftarrow{y}_\varepsilon(t) := e^{\int_0^t \mathbf{m}(x_\varepsilon(\sigma))d\sigma}\left(W(T,x) + \varepsilon - A(\overleftarrow{x}_\varepsilon(\cdot))(t)\right) \tag{6.51}$$

we infer that $\overleftarrow{y}_\varepsilon(0) = W(T,x)+\varepsilon$, that $\overleftarrow{y}_\varepsilon(\Omega^\flat) \geq \mathbf{c}(T-\Omega^\flat_\varepsilon, \overleftarrow{x}_\varepsilon(\Omega^\flat_\varepsilon))$ and that for all $t \in [\Omega^\flat_\varepsilon]$,

$$\begin{cases} (i)\ \overleftarrow{y}_\varepsilon(t) \geq \mathbf{k}(T-t, \overleftarrow{x}_\varepsilon(t)) \\ (ii)\ \overleftarrow{y}_\varepsilon(t) \geq -e^{\int_0^t \mathbf{m}(x_\varepsilon(\sigma))d\sigma} \left\langle \frac{\partial U(\overleftarrow{x}_\varepsilon(t))}{\partial x}, \overleftarrow{x}'_\varepsilon(t) \right\rangle \end{cases} \tag{6.52}$$

This means that $(x, T, W(T)+\varepsilon)$ belongs to the capture basin $\mathrm{Capt}_{(6.37)}(\mathscr{E}p(\mathbf{k}), \mathscr{E}p(\mathbf{c}))$. By definition of the viability solution, $V(T,x) \leq W(T,x)+\varepsilon$. Letting ε converge to 0 provides the inequality $V(T,x) \leq W(T,x)$, and thus, the equality $V(T,x) = W(T,x)$ we were looking for. ■

Theorem 6.3.6 (Viable and Optimal Evolutions Coïncide). *When $(T,x,W(T,x))$ belongs to the epigraph of the valuation function $(T,x) \mapsto W(T,x)$, then there exists at least one optimal evolution $\overleftarrow{x}_{\mathrm{opt}}(\cdot) \in \mathscr{S}_c(x)$ starting at time (T,x) at initial time 0 viable in $\mathscr{E}p(V)$ on the interval $[T-\Omega^\flat_{\mathrm{opt}}, T]$. The following conditions are equivalent:*

1. *The evolution $\overleftarrow{x}_{\mathrm{opt}}(\cdot) \in \mathscr{S}_c(T,x)$ defined on $[T-\Omega^\flat_{\mathrm{opt}}, T]$ is optimal;*
2. *The evolution $t \mapsto (T-t, \overleftarrow{x}_{\mathrm{opt}}(t), \overleftarrow{y}_{\overleftarrow{x}_{\mathrm{opt}}}(\cdot)(t))$ starting at $(T,x,W(T,x))$ at initial time is viable in the epigraph of W on $[T-\Omega^\flat_{\mathrm{opt}}, T]$.*

Proof. Optimal evolutions being viable, assume that $(T,x,W(T,x))$ belongs to the capture basin $\mathrm{Capt}_{(6.37)}(\mathscr{E}p(\mathbf{k}), \mathscr{E}p(\mathbf{c}))$ and associate with any $\overleftarrow{x}(\cdot) := \overleftarrow{x}_{\mathrm{opt}}(\cdot) \in \mathscr{S}_c(x)$ starting at x the evolution $\overleftarrow{y}(\cdot) := \overleftarrow{y}_{\mathrm{opt}}(\cdot)$ defined by

$$\overleftarrow{y}(t) := \overleftarrow{y}_{(\overleftarrow{x}_{\mathrm{opt}}(\cdot), W(T,x))}(t) \tag{6.53}$$

such that the evolution $t \mapsto (T-t, \overleftarrow{x}(t), \overleftarrow{y}(t))$ starting at $(0,x,W(T,x))$ is viable in $\mathscr{E}p(\mathbf{k})$ until time $\Omega^\flat_{\mathrm{opt}}$ when it reaches $\mathscr{E}p(\mathbf{c})$. Hence, we can take $\varepsilon = 0$ in formula (6.51), p. 101, and infer that

$$\begin{cases} W(T,x) = \max\left[e^{\int_0^{\Omega^\flat_{\mathrm{opt}}} \mathbf{m}(x(\sigma))d\sigma} \mathbf{c}(T-\Omega^\flat_{\mathrm{opt}}, x(\Omega^\flat_{\mathrm{opt}})) + A(\overleftarrow{x}(\cdot))(\Omega^\flat_{\mathrm{opt}}), \right. \\ \left. \sup_{t\in[0,\Omega^\flat_{\mathrm{opt}}]} \left(\max\left[\mathbf{k}(T-t, \overleftarrow{x}(t)), -\left\langle \frac{\partial U(\overleftarrow{x}(t))}{\partial x}, \overleftarrow{x}'(t) \right\rangle \right] + A(\overleftarrow{x}(\cdot))(t) \right) \right] \end{cases} \tag{6.54}$$

so that the evolution $\overleftarrow{x}(\cdot)$ achieves the minimum in $W(T,x)$, and thus is optimal. ■

6.3.3 Optimality Properties

We shall derive the dynamical optimality property of the valuation function $t \mapsto W(t, x(t))$ from the bilateral fixed point property of the capture basin defining the viability solution (see Theorem 38, p. 124).

Theorem 6.3.7 (Dynamic Optimality Property of the Valuation Function). *The valuation* $t \mapsto V(t, \overleftarrow{x}(t))$ *along an optimal (or viable) evolution satisfies the* dynamic optimality property*: Let* $\Omega_{\mathbf{k}}$ *be the first time such that* $V(T - \Omega_{\mathbf{k}}, \overleftarrow{x}(\Omega_{\mathbf{k}})) = \mathbf{k}(T - \Omega_{\mathbf{k}}, \overleftarrow{x}(\Omega_{\mathbf{k}}))$*. Then*

$$\forall\, t \in [0, \Omega_{\mathbf{k}}],\; e^{\int_0^t \mathbf{m}(x(\sigma))d\sigma} W(T - t, x(t)) + A(\overleftarrow{x}(\cdot))(t) \;=\; W(T, x) \quad (6.55)$$

Furthermore, $t \mapsto W(T - t, \overleftarrow{x}(t))$ *is the solution to the differential equation*

$$\begin{cases} \forall\, t \in [0, \Omega_{\mathbf{k}}],\; \dfrac{d}{dt} W(t, \overleftarrow{x}(t)) \;=\; -\mathbf{m}(\overleftarrow{x}(t)) \left(W(t, \overleftarrow{x}(t)) + \alpha U(\overleftarrow{x}(t)) \right) \\ -\mathbf{l}(-\overleftarrow{x}'(t)) - \alpha \left\langle \dfrac{\partial U(\overleftarrow{x}(t))}{\partial x}, \overleftarrow{x}'(t) \right\rangle \end{cases} \quad (6.56)$$

Proof. This is actually a property of the viability solution V, so that we use this notation instead of W in the proof. Indeed the Bilateral Fixed Point Theorem 8.2.5, p. 123, states that the capture basin $\mathscr{E}p(V) = \mathrm{Capt}_{(6.37)}(\mathscr{E}p(\mathbf{k}), \mathscr{E}p(\mathbf{c}))$ is the bilateral fixed point

$$\mathrm{Capt}_{(6.37)}(\mathscr{E}p(V), \mathscr{E}p(\mathbf{c})) \;=\; \mathscr{E}p(V) \;=\; \mathrm{Capt}_{(6.37)}(\mathscr{E}p(\mathbf{k}), \mathscr{E}p(V))$$

1. Since $(T, x, V(T, x)) \in \mathscr{E}p(V)$ belongs to the capture basin $\mathrm{Capt}_{(6.37)}(\mathscr{E}p(V), \mathscr{E}p(\mathbf{c}))$ of the epigraph of $\mathbf{c}$ viable in $\mathscr{E}p(V)$ under the auxiliary system (6.37), p. 99, inequality (6.47), p. 101, of the proof of Theorem 6.3.5, p. 99, where $\mathbf{k}$ is replaced by V, implies that

$$\forall\, t \in [0, \Omega^{\flat}],\; \left(e^{\int_0^t \mathbf{m}(x(\sigma))d\sigma} V(T - t, x(t)) + A(\overleftarrow{x}(\cdot))(t) \right) \;\leq\; V(T, x) \quad (6.57)$$

i.e., inequality

$$V(T - t, x(t)) \;\leq\; \overleftarrow{y}_{(\overleftarrow{x}(\cdot), V(T,x))}(t) \;:=\; e^{\int_0^t \mathbf{m}(x(\sigma))d\sigma} \left(V(T, x) - A(\overleftarrow{x}(\cdot))(t) \right) \quad (6.58)$$

2. We deduce the opposite inequality from the second fixed point property stating that $\mathscr{E}p(V) = \mathrm{Capt}_{(6.37)}(\mathscr{E}p(\mathbf{k}), \mathscr{E}p(V))$.

Since the viability solution is defined by $V(T,x) := \inf_{(T,x,y)\in \mathrm{Capt}_{(6.37)}(\mathscr{E}p(\mathbf{k}),\mathscr{E}p(V))} y$ and since we assumed that $(T,x,V(T,x))$ belongs to $\mathrm{Capt}_{(6.37)}(\mathscr{E}p(\mathbf{k}),\mathscr{E}p(V))$, whenever $V(T,x) > \overleftarrow{k}(0,x)$, we deduce that for any $\varepsilon > 0$ such that $\varepsilon < V(T,x) - \overleftarrow{k}(0,x)$, the triple $(T,x,V(T,x)-\varepsilon)$ does not belong to $\mathrm{Capt}_{(6.37)}(\mathscr{E}p(\mathbf{k}),\mathscr{E}p(V))$. By definition of capture basins, this implies that all evolutions $t \mapsto (T-t, \overleftarrow{x}(t), \overleftarrow{z}_\varepsilon(t))$ governed by the characteristic system (6.37), p. 99 starting at $(T,x,V(T,x)-\varepsilon)$ remain outside of the target $\mathscr{E}p(V)$ forever or until some $\Omega_{\mathbf{k}_\varepsilon} > 0$ when $\overleftarrow{z}_\varepsilon(\Omega_{\mathbf{k}_\varepsilon}) = \mathbf{k}(T-\Omega_{\mathbf{k}_\varepsilon}, \overleftarrow{x}(\Omega_{\mathbf{k}_\varepsilon}))$.

Take any viable evolution $\overleftarrow{x}(\cdot) \in \mathscr{S}_c(x)$ and associate with it

$$\overleftarrow{y}(t) := \overleftarrow{y}_{(\overleftarrow{x}(\cdot),V(T,x))}(t) := e^{\int_0^t \mathbf{m}(x(\sigma))d\sigma}\Big(V(T,x) - A(\overleftarrow{x}(\cdot))(t)\Big)$$

and

$$\overleftarrow{z}_\varepsilon(t) := \overleftarrow{y}(t) - \varepsilon e^{\int_0^t \mathbf{m}(x(\sigma))d\sigma} \tag{6.59}$$

Therefore, $z_\varepsilon(\cdot)$ is governed by the characteristic system (6.37), p. 99 starting at $(T,x,V(T,x)-\varepsilon)$, which remains outside the epigraph of V forever or until some $\Omega_{\mathbf{k}_\varepsilon} \in]0,T]$ when $z_\varepsilon(\Omega_{\mathbf{k}_\varepsilon}) = \overleftarrow{y}(\Omega_{\mathbf{k}_\varepsilon}) - \varepsilon e^{\int_0^{\Omega_{\mathbf{k}_\varepsilon}} \mathbf{m}(x(\sigma))d\sigma} = \mathbf{k}(T-\Omega_{\mathbf{k}_\varepsilon}, \overleftarrow{x}(\Omega_{\mathbf{k}_\varepsilon}))$.

This implies that

$$\forall\, t \in [0,\Omega_{\mathbf{k}_\varepsilon}],\ \overleftarrow{z_\varepsilon}(t) < V(T-t,\overleftarrow{x}(t)) \tag{6.60}$$

Let $\Omega_{\mathbf{k}}$ be the limit of some subsequence $\Omega_{\mathbf{k}_\varepsilon}$ when $\varepsilon \to 0+$, which satisfies $\overleftarrow{y}(\Omega_{\mathbf{k}}) = \mathbf{k}(T-\Omega_{\mathbf{k}}, \overleftarrow{x}(\Omega_{\mathbf{k}}))$. We deduce that

$$\forall\, t \in [0,\Omega_{\mathbf{k}}],\ \overleftarrow{y}(t) \le V(T-t,\overleftarrow{x}(t)) \tag{6.61}$$

and thus, using (6.58), p. 103, equality

$$\forall\, t \in [0,\Omega_{\mathbf{k}}],\ \overleftarrow{y}(t) = V(T-t,\overleftarrow{x}(t)) \tag{6.62}$$

Since $\overleftarrow{y}(\cdot)$ is the solution of the differential equation

$$\overleftarrow{y}'(t) = -\mathbf{m}(\overleftarrow{x}(t))\Big[\overleftarrow{y}(t) + \alpha U(\overleftarrow{x}(t))\Big] - \mathbf{l}(-\overleftarrow{x}'(t)) - \alpha\left\langle \frac{\partial U(\overleftarrow{x}(t))}{\partial x}, \overleftarrow{x}'(t)\right\rangle$$

starting at x, we infer property (7.18), p. 103. ■

Chapter 7
Endowing Exchange Values: Adam Smith's Invisible Man

7.1 A Mathematical Framework of Dynamical Economic Theory

Commodities, Prices, and Transaction Spaces

In economics, $X := \mathbb{R}^\ell$ can be regarded as a *commodity space* of commodities $x := (x_h)_{1 \le h \le \ell}$ of amounts $x_h \in \mathbb{R}$ of units e^h of goods or services labeled $h = 1, \ldots, \ell$.

Its dual $X^\star := \mathbb{R}^{\ell\star}$ is the space of *prices* $p : x \mapsto \langle p, x \rangle \in \mathbb{R}$ associating with any commodity $x \in X$ its *duality product*, regarded as a *value pairing* $\langle p, x \rangle := \sum_{h=1}^{\ell} p^h x_h$ which is interpreted as a *patrimonial value*. The values $p^h := \langle p, e^h \rangle \in \mathbb{R}$ of the units e^h are the components of the price $p \in X^\star$ in the dual basis.

The velocity $x'(t)$ at time t of the evolution of a commodity $x(\cdot)$ is regarded as a *transaction* (actually, an infinitesimal one). Transactions range over the bidual $X^{\star\star}$. In the same way, the *price fluctuations* $p'(t)$ range over the tridual of the $X^{\star\star\star}$.

There exists a canonical injection from X to its bidual $X^{\star\star}$, which is a bijection for finite dimensional spaces.

Even though these four spaces are all isomorphic, they play different roles in economics.[1]

---

[1]As well as in other domains, such that classical mechanics from which these ideas emerged: $X := \mathbb{R}^\ell$ can be regarded as a *position or configuration space*, its dual $X^\star$ as the space of *forces* $p : x \mapsto \langle p, x \rangle \in \mathbb{R}$ associating with any position $x \in X$ its *duality product* $\langle p, x \rangle$ which is interpreted as a *work*. The bidual $X^{\star\star}$ is regarded as the space of velocities v and the duality product $\langle p, v \rangle$ is interpreted as a *power*.

In his definition of happiness,[31–74] *Maupertuis* actually chose $X := \mathbb{R}$ as the space of durations d, its dual as the space of urges p, its bidual as the space of *fluidities* φ and its tridual as the space of hastes (see Sect. 5.3, p. 74).

J.-P. Aubin, *Time and Money*, Lecture Notes in Economics and Mathematical Systems 670, DOI 10.1007/978-3-319-00005-3_7,

In the case of duration functions, where $\mathbb{R}_+$ is the set of durations d, we propose to regard the dual variable $e \in \mathbb{R}$ as an *urge*. The product ed of the urge e and the duration d would be the *"precipitation"* and the product $p\varphi$ of the urge p and fluidity φ could be regarded as *"impetuosity."*

We study the evolution of the following (economic) variables:

$$x(t) \in X \ (commodities) \text{ and } p(t) \in X^\star \ (prices)$$

the derivatives of which are denoted by

$$x'(t) \in X^{\star\star} \ (transactions) \text{ and } p'(t) \in X^{\star\star\star} \ (price\ fluctuations) \tag{7.1}$$

In economics and finance, the *impetus* (or *value creation*) is the velocity $y'(t)$ of the *patrimonial value*, which is the sum of the transaction values $\langle p(t), x'(t)\rangle$ and fluctuation values $\langle p'(t), x(t)\rangle$.

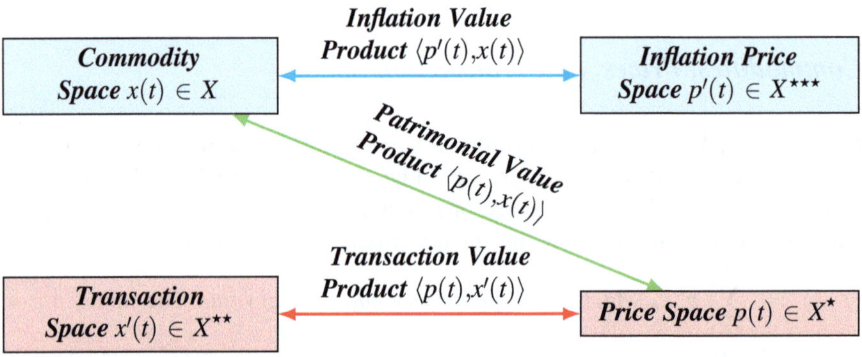

36 **[The Four Economic Interpretations of the Vector Space $\mathbb{R}^\ell$]**

The evolution of the commodities and prices and their velocities range over the four "economic spaces" symbolized in this scheme and we display the three duality products which are of use.

Economic Potential and Impetus

We consider a set of n (economic) agents. We denote by X^n the set of allocations $x := (x_i)_{i=1,\ldots,n}$ of commodities $x_i \in X$ among n agents. In order to determine temporal windows, we introduce the (vectorial) *duration function* $t \mapsto d(t) := (d_i(t))_{i=1,\ldots,n} \in \mathbb{R}^n_+$ of nondecreasing and nonnegative duration functions $d_i(\cdot)$ satisfying $d(T) = D$. Taking for duration function $d_i(t) := t$ with infinite aperture provides the case when the use of duration functions has no effect. Using the age durations $d_i(t) := t - (T - D_i)$ implies that the temporal windows under which evolve the commodities $x_i(t)$ are $[T - D_i, T]$, which adapt to the standard economic considerations the McKendrick approach of age-structured equations in mathematical demography and population dynamics (see [14, Aubin]).

Definition 7.1.1 (Economic Potentials and their Impetus). We introduce economic potential functions $U : (d, x, p) \in \mathbb{R}^n_+ \times X^n \times X^\star \mapsto U(d, x, p) \in \mathbb{R}$ associating with the economic "duration–allocation–price" triples (d, x, p) a number which as the dimension of an "economic value". We interpret their partial derivatives:

1. The partial derivatives $e_i(t, d; x, p) := \frac{\partial U(d, x, p)}{\partial d_i}$ are the individual *urges* denoting the instantaneous increase during an infinitesimal duration interval;
2. The partial derivatives $p_i(t, d; x, p) := \frac{\partial U(d, x, p)}{\partial x_i}$ are individual *prices* (as in marginal theory) denoting the instantaneous increase of value triggered by an infinitesimal variation of the commodity (instantaneous transaction);
3. The partial derivative $\frac{\partial U(d, x, p)}{\partial p} \in \mathbb{R}^\ell$ is the *global commodity*.

The *impetus* along an evolution $t \mapsto (d(t), x(t), p(t))$ is defined by

$$\begin{cases} E(d(t), x(t), p(t)) := \frac{d}{dt}U(d(t), x(t), p(t)) = \\ \sum_{i=1}^{n} \left(\left\langle \frac{\partial U(d(t); x(t), p(t))}{\partial d_i}, d_i'(t) \right\rangle + \left\langle \frac{\partial U(d(t); x(t), p(t))}{\partial x_i}, x_i'(t) \right\rangle \right) \\ + \left\langle \frac{\partial U(d(t); x(t), p(t))}{\partial p}, p'(t) \right\rangle \end{cases} \quad (7.2)$$

is the sum of the impetuosities (products of urges by durations), the transaction values, and the fluctuation impacts.

Since the word "value" is polysemous, we borrowed to the physicists the name "potential," which is a primitive concept providing the concept of force.[2] The economic potential function U is a primitive concept from which we derive other concepts, such that the endowment function, in the same way that utility functions are primitive concepts in static economics.

[2] In physics, the gradient of a *potential function* $U : x \mapsto U(x)$ is interpreted as a force: along an evolution $t \mapsto x(t)$, $\frac{d}{dt}U(x(t)) = \left\langle \frac{\partial U(x(t))}{\partial x}, x'(t) \right\rangle$ is a *power*. In economics, the variable x is replaced by the duration–allocation–price triple and the impetus plays the role of the mechanical power. For durations, the force is the urge and the impetuosity plays the role of power.

The canonical example of economic potential function involving duration, allocations of commodities and prices is the trilinear form

$$U(d,x,p) = \sum_{i=1}^{n} d_i \langle p, x_i \rangle \tag{7.3}$$

Hence

1. The individual *urges* are equal to $e_i(d,x,p) := \langle p, x_i \rangle$;
2. The individual *prices* are equal to $p_i(d,x,p) := d_i p$;
3. The *advised commodity* is equal to $\sum_{i=1}^{n} d_i x_i$;
4. The *impetus* is equal to

$$\begin{cases} E(d(t),x(t),p(t)) := \frac{d}{dt}U(d(t),x(t),p(t)) \\ = \sum_{i=1}^{n} \left(d_i'(t)\langle p(t), x_i(t)\rangle + \langle d_i(t)p(t), x_i'(t)\rangle\right) + \left\langle p(t), \sum_{i=1}^{n} d_i(t)x_i(t)\right\rangle \end{cases} \tag{7.4}$$

We denote Σ the summation operator

$$\forall\, x := (x_i)_{i=1,\dots,n} \in X^n, \quad \Sigma x := \sum_{i=1}^{n} x_i \in X$$

Its transpose $\Sigma^\star$ attributes to each economic agent the same prices $p \in X^\star$:

$$\forall\, p \in X^\star, \quad \Sigma^\star p := (p,\dots,p) \in X^{n\star} := (X^\star)^n$$

The impact of price fluctuation $\langle p'(t), x(t)\rangle$ on a commodity $x(t)$ is related to the concept of *inflation*, measured as the impact of price fluctuations $\langle p'(t), b\rangle$ on a numéraire or a *consumer price indexes* $b \in X$: $\langle p'(t), x(t)\rangle = \left(\frac{\langle p'(t), x(t)\rangle}{\langle p'(t), b\rangle}\right)\langle p'(t), b\rangle$.

We choose from now on the canonical economic potential function $U : (d,x,p) \mapsto U(d,x,p) = \sum_{i=1}^{n} d_i \langle p, x_i \rangle$.

Whenever there is no duration functions nor price fluctuations, then the impetus

$$E(d(\cdot),x(\cdot),p) := \sum_{i=1}^{n} \langle p, x_i'(t)\rangle \in \mathbb{R} \cup \{+\infty\} \tag{7.5}$$

boils down to the transaction values.

This is not a realistic assumption, since the prices are not constant, and produce price fluctuation impacts, and since duration functions may be involved to determine temporal windows. Mathematically, it costs nothing to assume that prices and duration functions evolve, except for the complexity of notations and length of formulas.

The balanced budgetary constraint requires that, on a temporal window, the maximal impetus should always be nonpositive:

$$\begin{cases} \sup_{t\in[T-\Omega(d(\cdot)),T]} E(d(\cdot),x(\cdot),p(\cdot)) \;:=\; \sup_{t\in[T-\Omega(d(\cdot)),T]} \\ \left(\sum_{i=1}^{n}(d_i'(t)\,\langle p(t),x_i(t)\rangle + \langle p(t),d_i(t)x_i'(t)\rangle + \langle p'(t),d_i(t)x_i(t)\rangle)\right) \;\leq\; 0 \end{cases} \tag{7.6}$$

This budgetary rule is the dynamical version of the static Walras law requiring that debts are forbidden at each instant.[3] Transactions, price fluctuations, duration, fluidities, and urges were not involved. In some cases, transactions are taken into account, but not price fluctuations.

By taking into account only the transaction values and forgetting the price fluctuation impact (inflations) and the impetuosity, the conclusions are different, since only nonpositive impetus is the assumption needed to guarantee the viability of the economic system described by commodities, prices, and durations.

Even though economic constraints are satisfied, such budgetary or financial constraints requiring that the impetus should be nonpositive may not be satisfied. *If one agrees that what matters is the viability of the duration–allocation–price triples, but not the sign of their impetus*, we have to compute the endowment necessary to insure that the transactions and price fluctuations are viable, without requiring that the impetus is always nonpositive.

One can have an idea of transactions, since they describe the behavior of the agents, but the laws of evolution of prices and values escape the power of the agents, and are delegated to the "Market." We then need to assume that "He," the invisible seer, knows what to do with His hand governing the evolution of prices and incomes, the other one guiding the transactions of the agents (actually, it is the Adam Smith man who is invisible, but with *two* hands, consequently invisible). However, the worshippers of this new deity forget the necessity of an endowment provider, a visible demiurge correcting the errors of the "Market".

---

[3]We could add constraints of the form

$$\begin{cases} C(d(\cdot),x(\cdot),p(\cdot)) \;:=\; \int_{T-\Omega(d(\cdot))}^{T} \\ \left(\sum_{i=1}^{n}(d_i'(t)\,\langle p(t),x_i(t)\rangle + \langle d_i(t)p(t),x_i'(t)\rangle + \langle p'(t),d_i(t)x_i(t)\rangle)\right) dt \;\leq\; 0 \end{cases} \tag{7.7}$$

on *cumulated* transactions and fluctuations on a temporal window, which are more classical since they involve integrals instead of suprema, which are difficult to deal with standard methods of calculus of variations. We observe that $C(d(\cdot),x(\cdot),p(\cdot)) \leq \Omega(d(\cdot)) \sup_{t\in[T-\Omega(d(\cdot)),T]} E(d(\cdot),x(\cdot),p(\cdot))$. It is also possible to add this cumulated cost to the stimulus (see Theorem 6.4.26, p. 232, of [15, *Viability Theory. New Directions*].).

Therefore, we are naturally enticed to study the relations between the *joint evolution of commodities, prices, and durations* for the economical constraints to be viable, but not necessarily financially balanced.

Economic Constraints

We denote by $(t,d) \in \mathbb{R} \times \mathbb{R}_+ \rightsquigarrow K(t,d) \subset X^n \times X^\star$ the *time–duration map* describing the "economic environment" $K(t,d) \subset X^n \times X^\star$ of available and desired commodities and prices. *We require that chaperoned evolutions* $(x(\cdot), p(\cdot))$ *are* viable *in the associated tube* $t \in \mathbb{R} \rightsquigarrow K(t,d(t)) \subset X^n \times X^\star$ *on the temporal window* $[T-\Omega(d(\cdot)),T]$ in the sense that for any current time $t \in [T-\Omega(d(\cdot)),T]$, $(x(\cdot), p(\cdot))$ belongs to $K(t,d(t))$.

The standard example is the set

$$\begin{cases} K(t,d) := \{x \in \prod_{i=1}^n L_i(t,d) \mid \sum_{i=1}^n x_i \in M(t,d) \subset Y \times P(t,d) \\ \text{and } \forall\, i = 1,\ldots,n,\ \langle p, x_i \rangle \geq z_i(t,d)\} \end{cases} \tag{7.8}$$

of duration functions $d_i \in \mathbb{R}_+$ of allocations $x_i \in L_i(t,d)$ (viability constraints of agents) of commodities $y \in M(t,d)$ (scarcity constraints) among the agents, of prices $p \in P(t,d) \subset Y^\star$ (for instance, $P := \mathbb{R}^\ell_+$) and wealth thresholds $z_i(t,d)$, most of the time assumed to be nonnegative for consumer and nonpositive for producers.

We can associate with economic constraints several *"constraint cost functions,"* among which

1. the constraint function $\mathbf{k}_{-\infty}$ defined by

$$\mathbf{k}_{-\infty}(t,d,x) := \begin{cases} -\infty \text{ if } x \in K(t,d) \\ +\infty \text{ if } x \notin K(t,d) \end{cases} \tag{7.9}$$

(a negative cost is a profit, and no constraints are imposed on profits);

2. the constraint function $\mathbf{k}_0$ defined by

$$\mathbf{k}_0(t,d,x) := \begin{cases} 0 \quad \text{if } x \in K(t,d) \\ +\infty \text{ if } x \notin K(t,d) \end{cases} \tag{7.10}$$

which imposes a cost, equal to 0, whenever the commodity obeys the constraints;

3. or any cost function[4] $(t,d,x,p) \mapsto \mathbf{k}(t,d,x,p)$ satisfying

$$\forall\, x \notin K(t,d),\ \mathbf{k}(t,d,x,p) = +\infty \tag{7.11}$$

---

[4] Which can be regarded a kind of *fuzzy economy* in the sense of fuzzy or toll-sets. See [22, Aubin & Dordan].

Observe that if a constraint cost function $\mathbf{k} : (t, d, x, p) \mapsto \mathbf{k}(t, d, x, p) \in \mathbb{R} \cup \{+\infty\}$ is given, it defines a time–duration map

$$(t, d) \rightsquigarrow K(t, d) := \{(x, p) \in X^n \times X^\star \text{ such that } \mathbf{k}(t, d, x, p) < +\infty\}$$

See Chaps. 13, p. 523, 15, p. 603 and 17, p. 681, of [15, *Viability Theory. New Directions*] for more details.

Instantaneous Costs

An *instantaneous cost function* $(T, x, p) \mapsto \mathbf{c}(T, x, p)$ of the state $(x, p) \in X^n \times X^\star$ at instant $T := [T, T]$ (of zero aperture) is a function which associates the cost of x at *instant* T, *regarded as a temporal window with zero aperture.*

We shall assume once and for all that this cost function is *lower semicontinuous*, i.e., that its epigraph $\mathscr{E}p(\mathbf{c})$ is closed.

Economic Dynamics

For simplicity of the exposition, and also for realistically take into account our ignorance of the precise dynamical behavior of the agents and of the "Market," we shall assume only that we know some bounds on the agents' fluidities of time and transactions of commodities, and on the "Market's" price fluctuations.

Actually, the point of this study is precisely to deduce from economical and financial constraints the consistent dynamical behavior of the agents and the "Market" for insuring the economic viability: these are the transaction rules of the agent, the adjustment rules of the "Market," and the urge rules.

Definition 7.1.2 (Canonical a priori Bounded Dynamics). From now on, we shall use only *canonicals dynamics* described by a priori dynamical behaviors are described by bounds:

1. bounds $0 \leq \varphi_i^\flat(t) \leq \varphi_i^\sharp(t) < +\infty$ on the fluidities of the duration functions;
2. bound $0 \leq c_i(t) < +\infty$ on the norm of the commodity transactions of the agents i, $i = 1, \ldots, n$;
3. bound $0 \leq c_0(t) < +\infty$ on the norm of the price fluctuations of the "Market".

We denote by $\mathscr{A}(K(\cdot); T, D; x, p) := \mathscr{A}_{(\varphi_i^\flat, (\varphi_i^\sharp, c_i, c_0))}(K(\cdot); T, D; x, p)$ the set of evolutions $(d(\cdot), x(\cdot), p(\cdot))$ of evolutions $t \mapsto d(t) := (d_i(t))_{i=1,\ldots,n}$, $t \mapsto x(t) := (x_i(t))_{i=1,\ldots,n}$ and $t \mapsto p(t)$ with bounded velocities, *economically viable* in the sense that $\forall\ t \leq T, \ \ (x(t), p(t)) \in K(t, d(t))$ and arriving at (D, x, p) at terminal time T.

The canonical dynamical system requires only the knowledge of the time–duration constraint map $(t,d) \rightsquigarrow K(t,d)$ *and of the bounds* $\varphi^\flat(t)$, $\varphi^\sharp(t)$ and $c(t)$. This will be our favorite example.

For simplicity of the notations, we shall take constant velocity bounds and assume that the return $\mathbf{m}(p)$ depends only on the price $p \in X^\star$ instead of $\mathbf{m}(d,x,p)$, whereas the formulas allow the dependence of the return on durations and commodities. More information on the dynamical behavior, if available, can be taken into account in other examples.

7.2 Endowment Functions

Potential functions are given as a priori primitive data *independently of the economic constraints*, so that they are not specifically designed to guarantee the viability of economic evolutions under economic and financial constraints, introduced independently of them.

For taking them into account, we shall correct such potential functions by introducing *a posteriori* or *ex post* endowment functions $W : (T,D;x,p) \mapsto W(T,D;x,p)$ hedging the impetus of viable duration–commodity–transaction triples.

Definition 7.2.1 (The Endowment Function). The economic constraints $(x,p) \in K(t,d)$, the instantaneous cost $\mathbf{c}$, the bounds of the canonical dynamic system, and the return map $\mathbf{m}$ being given, the endowment function $W : (T,D;x,p) \mapsto (T,D;x,p)$ is defined by

$$\begin{cases} W(T,D;x,p) := \inf_{(d(\cdot),x(\cdot),p(\cdot))\in\mathscr{A}(K(\cdot);T,D;x,p)} \inf_{\Omega\geq 0} \\ \max\left[e^{\int_{T-\Omega}^{T}\mathbf{m}(p(\sigma))d\sigma}\mathbf{c}(T-\Omega,x(T-\Omega),p(T-\Omega)),\right. \\ \left.\sup_{t\in[T-\Omega,T]} e^{\int_t^T \mathbf{m}(p(\sigma))d\sigma} E(d(t),x(t),p(t))\right] \end{cases} \tag{7.12}$$

which is the infimum over the set of economically viable evolutions $(d(\cdot),x(\cdot),p(\cdot))$ of their maximal actualized impetus.

The endowment function $W(T,D;x,p)$ involves the maximum of

1. the *instantaneous functional*

$$\mathbf{J_c}(T) := e^{\int_{T-\Omega}^{T}\mathbf{m}(p(\sigma))d\sigma}\mathbf{c}(T-\Omega,x(T-\Omega),p(T-\Omega)) \tag{7.13}$$

depending on the instantaneous cost function $\mathbf{c}(\cdot)$;

2. the *maximal impetus functional*

$$\mathbf{I}_U(T) := \sup_{t\in[T-\Omega,T]} e^{\int_t^T \mathbf{m}(p(\sigma))d\sigma} E(d(t),x(t),p(t)) \tag{7.14}$$

depending on the economic potential $U(d, x, p)$.

Note that the subsets

1. $\{(T, D; x, p)$ such that $W(T, D; x, p) := +\infty\}$ is the subset of states $x \in K(T)$ which are neither economically nor financially viable;
2. $\{(T, D; x, p)$ such that $0 < W(T, D; x, p) < +\infty\}$ is the subset of states $x \in K(T)$ which are economically viable but not financially viable;
3. $\{(T, D; x, p)$ such that $-\infty < W(T, D; x, p) \leq 0\}$ is the subset of states $x \in K(T)$ which are both economically and financially viable.

When chronological time and duration do evolve, we denote them by lower case t and d.

Definition 7.2.2 (Partial Derivatives of an Endowment Function). We interpret the partial derivatives of the endowment function in the following way:

1. The partial derivative $\frac{\partial W(t, d; x, p)}{\partial t}$ is a *flux* denoting the instantaneous increase during an infinitesimal time interval (the "Greek Θ" in finance, or *time decay*);
2. The partial derivatives $e_i(t, d; x, p) := \frac{\partial W(t, d; x, p)}{\partial d_i}$ are the individual *urges* denoting the instantaneous increase during an infinitesimal duration interval;
3. The partial derivatives $p_i(t, d; x, p) := \frac{\partial W(t, d; x, p)}{\partial x_i}$ are individual *prices* (as in marginal theory) denoting the instantaneous increase of value triggered by an infinitesimal variation of the commodity (instantaneous transaction);
4. The partial derivative $\frac{\partial W(t, d; x, p)}{\partial p} \in \mathbb{R}^\ell$, the *advised global commodity* (the "Greek Δ" in finance).

We observe that, for any evolution $t \mapsto (d(t), x(t), p(t))$, the instantaneous increase of endowment is the sum

$$
\begin{cases}
\frac{d}{dt} W(t, d(t); x(t), p(t)) = \frac{\partial W(t, d(t); x(t), p(t))}{\partial t} \\
+ \sum_{i=1}^{n} \left(\left\langle \frac{\partial W(t, d(t); x(t), p(t))}{\partial d_i}, d_i'(t) \right\rangle + \left\langle \frac{\partial W(t, d(t); x(t), p(t))}{\partial x_i}, x_i'(t) \right\rangle \right) \\
+ \left\langle \frac{\partial W(t, d(t); x(t), p(t))}{\partial p}, p'(t) \right\rangle
\end{cases}
\tag{7.15}
$$

of the flux, the impetuosities (products of urges by durations), the transaction values, and the price fluctuation impacts.

The endowment function $W(t,d,x,p)$ provides the agent the *derived* individual urges $\frac{\partial W(t,d(t),x(t),p(t))}{\partial d_i}$, individual prices $\frac{\partial W(t,d(t),x(t),p(t))}{\partial x_i}$ and the "Market" the derived global commodity $\frac{\partial W(t,d(t),x(t),p(t))}{\partial p}$, which *correct* the *advised* individual urges $\langle p(t), x_i'(t)\rangle$, individual prices $d_i(t)p(t)$ and the "Market" the advised global commodity $\sum_{i=1}^{n} d_i(t)x_i(t)$ from the canonical potential function U involved in the construction of the endowment function.

Remark (Financial Interpretation). In mathematical finances, the commodities $x(t) := P(t)$ are the shares of assets constituting a portfolio and $p(t) := S(t)$ are the prices of the assets, the value of which is $w(t) := \langle P(t), S(t)\rangle$. A peculiarity of mathematical finance approach is to ignore the mechanism of price formation, and to rather assume that the evolution of prices is uncertain. This is what is done in stochastic finance, when the dynamics of the prices is assumed to be stochastic ($dS(t) = r(S(t))S(t)dt + \sigma(S(t))S(t)dB(t)$) where dB is a Brownian movement, in order to overcome the ignorance of the price fluctuation, replacing its evolution by a stochastic process. However, we know at each date that the prices evolve in the "price tube" $t \mapsto \Sigma(t) := [S^{\flat}(t), S^{\sharp}(t)]$ (low and high bounds on prices), from which we derive that their velocities range over the (graphical) derivative $D\Sigma(t, S(t))$. Unfortunately, *the price tube is not viable under stochastic differential equations*, which can replace a priori stochastic differential equations involving volatilities $\sigma(S(t))$, furthermore difficult to measure directly for lack of a "volatilimeter." However, these flaws can be overcome by an alternative mathematical traduction of the concept of uncertainty, *tychastic* uncertainty (see [19, Aubin, Chen Luxi, Dordan & Saint-Pierre], Chap. 19 of [18, Aubin, Chen & Dordan] and [20, Aubin, Chen, Dordan, Faleh, Lezan & Planchet]) and [17, *Tychastic Measure of Viability Risk. A Viabilist Portfolio Performance and Insurance Approach*]. ■

7.3 Optimality Theorems on Endowment Functions

We state the main result deduced from Theorems 6.3.2, p. 98 and 6.3.3, p. 98:

Theorem 7.3.1 (Optimal Evolutions). *There exists at least one optimal economically evolution $(d_{\text{opt}}(\cdot), x_{\text{opt}}(\cdot), p_{\text{opt}}(\cdot)) \in \mathscr{A}(K(\cdot); T, D, x, p)$ arriving (D, x, p) at time T. We set $\Omega_{\text{opt}} := \max_{i=1,\dots,n} \Omega(d_{\text{opt}_i}(\cdot))$. The following conditions are equivalent:*

1. *The evolution $(d_{\text{opt}}, x_{\text{opt}}(\cdot), p_{\text{opt}}(\cdot)) \in \mathscr{A}(K(\cdot); T, D; x, p)$ is optimal in the sense that, $\forall\, t \in [T - \Omega_{\text{opt}}, T]$,*

$$W(T, D; x, p) = \sup_{t \in [T-\Omega_{\text{opt}}, T]} \left(e^{\int_t^T \mathbf{m}(p_{\text{opt}}(\sigma))d\sigma} E(d_{\text{opt}}(t), x_{\text{opt}}(t), p_{\text{opt}}(t)) \right) \tag{7.16}$$

2. *The evolution* $(d_{\text{opt}}, x_{\text{opt}}(\cdot), p_{\text{opt}}(\cdot)) \in \mathscr{A}(K(\cdot); T, D; x, p)$ *satisfies the* dynamic optimality *along the optimal evolution: whenever* $W(T, D; x, p) \in \mathbb{R}$ *is finite,* $\forall\, t \in [T - \Omega_{\text{opt}}, T]$,

$$e^{-\int_t^T \mathbf{m}(p_{\text{opt}}(\sigma))d\sigma} W(t, d_{\text{opt}}(t), x_{\text{opt}}(t), p_{\text{opt}}(t)) = W(T, D; x, p) \tag{7.17}$$

Furthermore, the endowment function $t \mapsto W(t, d_{\text{opt}}(t), x_{\text{opt}}(t), p_{\text{opt}}(t))$ *is the solution to the differential equation:* $\forall\, t \in [T - \Omega_{\text{opt}}, T]$,

$$\frac{d}{dt} W(t, d_{\text{opt}}(t), x_{\text{opt}}(t), p_{\text{opt}}(t)) = \mathbf{m}(p(t))(W(t, d_{\text{opt}}(t), x_{\text{opt}}(t), p_{\text{opt}}(t))) \tag{7.18}$$

satisfying the terminal condition $W(T, d_{\text{opt}}(T), x_{\text{opt}}(T), p_{\text{opt}}(T)) = W(T, D; x, p)$.

7.4 Microscopic and Macroscopic Properties

We adapt to the endowment function the two complementary mathematical properties of the endowment function obtained in the case of fundamental values.

We introduce the set-valued map $\mathscr{U}$ defined by

$$\begin{cases} \mathscr{U}(d, x, p, y) := \left\{ (\varphi, \upsilon, \pi) \in \left(\prod_{i=1}^{n} [\varphi_i^\flat, \varphi_i^\sharp(t, d)] \times c_i B \right) \times c_0 B_\star \right. \\ \left. \text{such that } \sum_{i=1}^{n} (\varphi_i \langle p, x_i \rangle + d_i(\langle p, \upsilon_i(t) \rangle + \langle \pi, x_i \rangle)) \le y \right\} \end{cases} \tag{7.19}$$

and set the gradients of the potential and endowment functions

$$\begin{cases} DU(d, x, p) := \left((\langle p, x_i \rangle)_{i=1,\ldots,n}, (d_i p)_{i=1,\ldots,n}, \sum_{i=1}^{n} d_i x_i \right) \\ DW(t, d, x, p) := \\ \left(\left(\frac{\partial W(t, d; x, p)}{\partial d_i} \right)_{i=1,\ldots,n}, \left(\frac{\partial W(t, d; x, p)}{\partial x_i} \right)_{i=1,\ldots,n}, \frac{\partial W(t, d; x, p)}{\partial p} \right) \end{cases} \tag{7.20}$$

and

$$\begin{cases} \langle DW(t,d,x,p),(\varphi,\upsilon,\pi)\rangle := \\ \sum_{i=1}^{n}\left(\left\langle\frac{\partial W(t,d;x,p)}{\partial d_i},\varphi_i\right\rangle+\left\langle\frac{\partial W(t,d;x,p)}{\partial x_i},\upsilon_i\right\rangle\right)+\left\langle\frac{\partial W(t,d;x,p)}{\partial p},\pi\right\rangle \end{cases} \tag{7.21}$$

Next, we define

1. the *"Hamiltonian" function* $\mathbf{l}^\star$ defined by

$$\begin{cases} \mathbf{l}_c^\star\,(DU(d,x,p),DW(t,d,x,p),W(t,d,x,p)) := \\ \inf_{(\varphi,\upsilon,\pi)\in\mathscr{U}(d,x,p,W(t,d,x,p))}\langle DW(t,d,x,p),(\varphi,\upsilon,\pi)\rangle \end{cases} \tag{7.22}$$

2. the (possibly empty) set-valued *regulation map* R defined by

$$\begin{cases} R\,(DU(d,x,p),DW(t,d,x,p),W(t,d,x,p)) := \\ \{(\varphi,\delta,\pi)\in\mathscr{U}(d,x,p,W(t,d,x,p)) \text{ such that} \\ \langle DW(t,d,x,p),(\varphi,\upsilon,\pi)\rangle = \mathbf{l}_c^\star\,(DU(d,x,p),DW(t,d,x,p)W(t,d,x,p))\} \end{cases} \tag{7.23}$$

The Hamiltonian function $\mathbf{l}^\star$ and the regulation map R depend on the endowment function $W(t,d,x,p)$, its urges $\frac{\partial W(t,d,x,p)}{\partial d_i}$, its (individual) prices $\frac{\partial W(t,d,x,p)}{\partial x_i}$, and its global commodity $\frac{\partial W(t,d,x,p)}{\partial p}$. They "correct" the given (or advised) urges $(\langle p,x_i\rangle)_{i=1,\ldots,n}$, the (individual) prices $(d_i\,p)_{i=1,\ldots,n}$, and the global commodity $\sum_{i=1}^{n} d_i x_i$ obtained from the datum of the potential function $U(d,x,p) := \sum_{i=1}^{n} d_i\,\langle p,x_i\rangle$ used as a potential function.

Knowing the endowment function $(t,x)\mapsto W(t,d,x,p)$ and assuming that it is continuously differentiable, we obtain

1. The *Microscopic Property*: Optimal evolutions $x(\cdot) := x_{\mathrm{opt}}(\cdot)$ are governed by the differential inclusion: $\forall\, t\in[T-\Omega_{\mathrm{opt}},T]$,

$$\begin{cases} \forall\, t\in[T-\Omega,T],\;\; (d'(t),x'(t),p(t)) \in \\ R\,(DU(d(t),x(t),p(t)),DW(t,d(t),x(t),p(t)),W(t,d(t),x(t),p(t))) \end{cases} \tag{7.24}$$

   satisfying the terminal condition $D(T)=T$, $x(T)=x$ and $p(T)=p$.
2. The *Macroscopic Property*: The endowment function $W(t,d,x,p)$ is a solution to the Hamilton–Jacobi equation

$$\begin{cases} \dfrac{\partial W(t,d,x,p)}{\partial t} = \\ \mathbf{l}_c^\star\,(DU(d,x,p), DW(t,d,x,p), W(t,d,x,p)) + \mathbf{m}(x)W(t,d,x,p) \end{cases} \tag{7.25}$$

satisfying the *instantaneous condition* $W(t,0,x,p) = \mathbf{c}(t,x,p)$.

The *regulation law* (7.24), p. 116 is a *"viable bilateral and optimal tâtonnement"*:

1. ***viable*** because it *governs the evolution of viable economic states and prices* in the sense that, for every $t \in [T - \Omega, T]$, $(x(t), p(t)) \in K(t, d(t))$;
2. ***bilateral*** because it governs both the evolution of the commodity and price, as well as the durations, and not only the evolution of prices, as the Walras tâtonnement does;
3. ***tâtonnement*** because duration functions, the transitions, and the price fluctuation depend on the discrepancies between the advised urges, individual prices, and total commodity on one hand, and on the advised urges, individual prices, and commodity defined by the endowment function, on the other hand;
4. ***optimal*** because the evolutions governed by the viable bilateral tâtonnement (7.24), p. 116, *optimize the endowment function on its graph.*

Chapter 8
Why Viability Theory?

8.1 Viability Kernels and Capture Basins

Viability theory ([7, 9, Aubin] and [15, Aubin, Bayen & Saint-Pierre]) designs and develops mathematical and algorithmic methods for investigating the *adaptation to viability constraints of evolutions governed by complex systems under uncertainty* that are found in many domains involving living beings, from biological evolution to economics, from environmental sciences to financial markets, from control theory and robotics to cognitive sciences.

It deals with the confrontation between viability and/or scarcity constraints and a controlled or regulated system $x'(t) = f(x(t), u(t))$ parameterized by controls or regulons (regulatory controls) $u(t) \in U(t, x(t))$, where $U : (t, x) \rightsquigarrow U(t, x)$ is the *contingent map*. The subset $U(t, x)$ describes a *"contingent reservoir"* of available controls or regulons for finding velocities guiding an evolution $x(\cdot)$ satisfying some required properties, for instance, viability in an environment and/or capturability of a target. The larger the contingent reservoir, the more *redondant* the system for regulating viable evolutions.

We denote by $\mathcal{S}(x)$ the subset of evolutions starting at x and governed by the regulated system

$$x'(t) \ = \ f(x(t), u(t)) \text{ where } u(t) \in U(x(t)) \tag{8.1}$$

The set-valued map $\mathcal{S} : x \rightsquigarrow \mathcal{S}(x)$ is called the *evolutionary system* generated by the regulated system.

Definition 8.1.1 (Viability Kernel and Capture Basin). Let $K \subset X$ be a environment and $C \subset K$ be a target.

1. The subset $\text{Viab}_{\mathcal{S}}(K, C)$ of initial states $x_0 \in K$ such that **at least one evolution** $x(\cdot) \in \mathcal{S}(x_0)$ starting at x_0 is viable in K for all $t \geq 0$ or viable

J.-P. Aubin, *Time and Money*, Lecture Notes in Economics and Mathematical Systems 670, DOI 10.1007/978-3-319-00005-3_8,

in K until it reaches C in finite time is called the *viability kernel of K with target C* under $\mathcal{S}$.

When the target $C = \emptyset$ is the empty set, we say that $\text{Viab}_{\mathcal{S}}(K) := \text{Viab}_{\mathcal{S}}(K, \emptyset)$ is the *viability kernel* of K. We set $\text{Capt}_{\mathcal{S}}(K, \emptyset) = \emptyset$.

2. The subset $\text{Capt}_{\mathcal{S}}(K, C)$ of initial states $x_0 \in K$ such that **at least one evolution** $x(\cdot) \in \mathcal{S}(x_0)$ starting at x_0 is viable in K until it reaches C in finite time is called the *capture basin of C viable in K* under $\mathcal{S}$. When $K = X$ is the whole space, we say that $\text{Capt}_{\mathcal{S}}(C) := \text{Capt}_{\mathcal{S}}(X, C)$ is the *capture basin* of C.

We say that

1. a subset K is *viable* under $\mathcal{S}$ if $K = \text{Viab}_{\mathcal{S}}(K)$,
2. K is *viable outside the target* $C \subset K$ under the evolutionary system $\mathcal{S}$ if $K = \text{Viab}_{\mathcal{S}}(K, C)$,
3. C is *isolated* in K if $C = \text{Viab}_{\mathcal{S}}(K, C)$,
4. K is a *repeller* if $\text{Viab}_{\mathcal{S}}(K) = \emptyset$, i.e., if the empty set is isolated in K.

We observe that *whenever $K \setminus C$ is a repeller, the capture basin of a target coincides with the viability kernel with this target:* $\text{Viab}_{\mathcal{S}}(K, C) = \text{Capt}_{\mathcal{S}}(K, C)$.

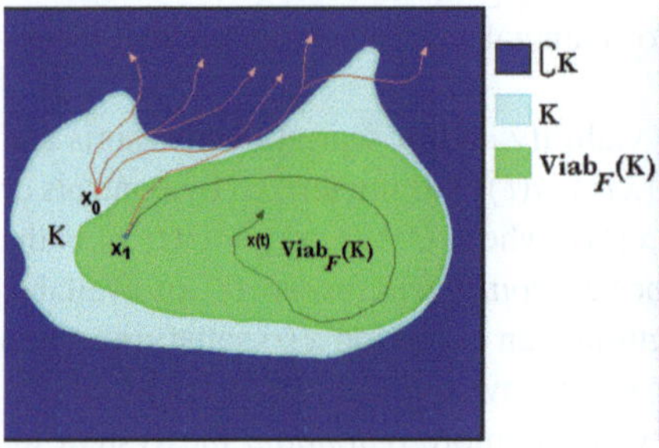

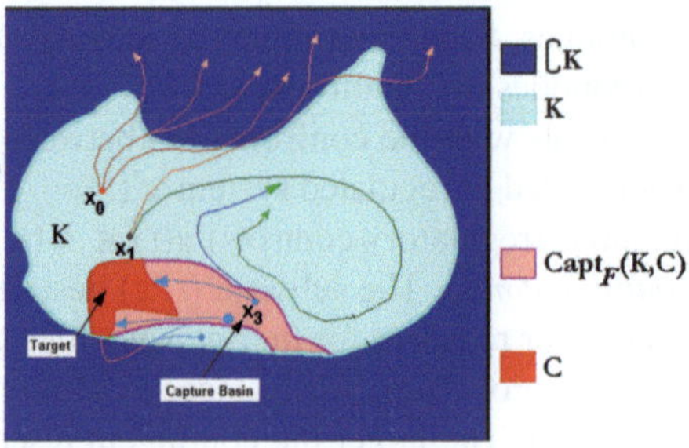

In control theory, open and closed loop controls, feedbacks or retroactions provide the central concepts of cybernetics and general systems theory:

Definition 8.1.2 (Retroactions). Retroactions are single-valued maps $\tilde{u} : (t, x) \in \mathbb{R}_+ \times X \mapsto \tilde{u}(t, x) \in \mathcal{U}$ that are plugged as inputs in the differential equation

$$x'(t) = f(x(t), \tilde{u}(t, x(t))) \tag{8.2}$$

In control theory, state-independent retroactions $t \mapsto \tilde{u}(t, x) := u(t)$ are called *open loop controls* whereas time-independent retroactions $x \mapsto \tilde{u}(t, x) := \tilde{u}(x)$ are called *closed loop controls* or *feedbacks*.

The class $\tilde{\mathcal{U}}$ in which retroactions are taken must be consistent with the properties of the parameterized system so that

- the differential equations $x'(t) = f(x(t), \tilde{u}(t, x(t)))$ have solutions,[1]
- for every $t \geq 0$, $\tilde{u}(t, x) \in U(x)$.

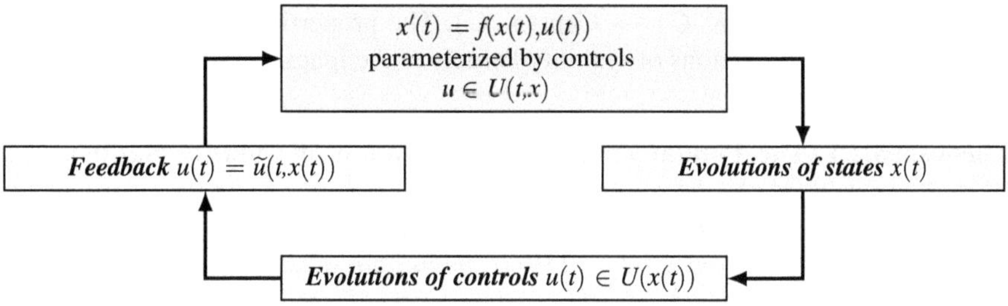

37 (Closed Loop Systems) *This diagram illustrates the regulated tychastic system mapping evolutions of controls to evolution of states and the feedback.*

The solution to this problem is given *in terms of subsets*: the viability kernel $\text{Viab}(K)$ and the graph of the retroaction map (graphical approach of maps) and uses the tools of set-valued analysis and mutational analysis (see [11, *Mutational and morphological analysis: tools for shape regulation and morphogenesis*] for instance). They are not obtained through analytical formulas, but can be computed in the framework of "set-valued numerical analysis." The viability algorithms and software handle at each iteration subsets instead of vectors.

8.2 A Viability Survival Kit

We summarize here some of the results of [15, Aubin, Bayen & Saint-Pierre] used in this study. This section presents few selected statements that are most often used, restricted to viable-capture basins only. Three categories of statements are presented. The first one provides characterizations of viable-capture basins as bilateral fixed points, which are simple, important, and valid without any assumption. The second one provides characterizations in terms of local viability properties and backward invariance, involving topological assumptions on the evolutionary systems. The third one characterizes viability kernels and viable-capture basins under differential inclusions in terms of tangential conditions, which furnishes the regulation map allowing to pilot viable evolutions (and optimal evolutions in the case of optimal control problems).

[1]Open loop controls can be only measurable. When the map $f : X \times \mathcal{U} \mapsto X$ is continuous, the Carathéodory Theorem states that differential equation (8.1), p. 119 has solutions even when the open loop control is just measurable, (and thus, can be discontinuous as a function of time). This is the case whenever they take their values in a finite set, in which case they are usually called "bang-bang controls" in the control terminology.

8.2.1 Bilateral Fixed Point Characterization

We consider the map $(K, C) \mapsto \text{Capt}(K, C)$. The properties of this map provide fixed point characterizations of viability kernels of the maps $K \mapsto \text{Capt}(K, C)$ and $C \mapsto \text{Capt}(K, C)$:

Theorem 8.2.1 (The Fundamental Characterization of Capture Basins). *Let $\mathcal{S} : X \rightsquigarrow \mathcal{C}(0, \infty; X)$ be an evolutionary system, $K \subset X$ be an environment, and $C \subset K$ be a nonempty target. The viable-capture basin $\text{Capt}_{\mathcal{S}}(K, C)$ of C viable in K (see Definition 8.1.1, p. 119) is the* ***unique*** *subset D between C and K that is both*

1. ***viable outside*** *C (and is the largest subset $D \subset K$ viable outside C),*
2. *satisfying $\text{Capt}_{\mathcal{S}}(K, C) = \text{Capt}_{\mathcal{S}}(K, \text{Capt}_{\mathcal{S}}(K, C))$ (and is the smallest subset $D \supset C$ to do so):*

i.e., the bilateral fixed point

$$\text{Capt}_{\mathcal{S}}(\text{Capt}_{\mathcal{S}}(K, C), C) = \text{Capt}_{\mathcal{S}}(K, C) = \text{Capt}_{\mathcal{S}}(K, \text{Capt}_{\mathcal{S}}(K, C)) \quad (8.3)$$

8.2.2 Viability Characterization

From now on, we assume that the evolutionary system is Marchaud:

Definition 8.2.2 (Marchaud Systems). We say that

1. a single-valued map $(x, u) \mapsto f(x, u) \in X$ is Marchaud if

$$\begin{cases} (i) & f \text{ is continuous} \\ (ii) & \text{for any } x,\ u \mapsto f(x, u) \text{ is affine} \\ (iii) & \exists\, c_0 > 0 \text{ such that } \forall x \in X,\ \|f(x, u)\| \le c_0(\|x\| + 1) \end{cases} \quad (8.4)$$

2. a set-valued map U is a *Marchaud map* if

$$\begin{cases} (i) & \text{the graph and the domain of } U \text{ are nonempty and closed} \\ (ii) & \text{the values } U(x) \text{ of } U \text{ are convex} \\ (iii) & \exists\, c > 0 \text{ such that } \forall x \in X,\ \|U(x)\| := \sup_{u \in U(x)} \|u\| \le c(\|x\| + 1) \end{cases} \quad (8.5)$$

3. the control system (f, U) defined by (8.1), p. 119 is Marchaud if both f and U are Marchaud.

All properties on viability kernels and capture basins require that the system is Marchaud. The canonical systems defined by Definitions 6.2.1, p. 87, and 7.1.2, p. 111, are Marchaud.

It happens that isolated subsets are, under adequate assumptions, backward invariant. Characterizing viability kernels and viable-capture basins in terms of forward viability and backward invariance allows us to use the results on viability and invariance.

Definition 8.2.3 (Local Viability and Backward Relative Invariance). A subset K is said *locally viable* under $\mathcal{S}$ if from any initial state $x \in K$ *there exist at least one evolution* $x(\cdot) \in \mathcal{S}(x)$ and a strictly positive $T_{x(\cdot)} > 0$ such that $x(\cdot)$ is viable in K on the nonempty interval $[0, T_{x(\cdot)}]$. It is a *repeller* under F if all solutions starting from K leave K in finite time.

A subset D is *backward invariant relatively to* K if all backward solutions starting from D viable in K are actually viable in K.

If K is itself backward invariant, any subset backward invariant relatively to K is backward invariant. If $C \subset K$ is backward invariant relatively to K, then $C \cap \mathrm{Int}(K)$ is backward invariant.

Proposition 8.2.4 (Capture Basins of Relatively Invariant Targets). *Let $C \subset D \subset K$ three subsets of X.*

1. *If D is backward invariant relatively to K, then* $\mathrm{Capt}_{\mathcal{S}}(K, C) = \mathrm{Capt}_{\mathcal{S}}(D, C)$,
2. *If C is backward invariant relatively to K, then* $\mathrm{Capt}_{\mathcal{S}}(K, C) = C$.

Using the concept of backward invariance, we provide a further characterization of viable-capture basins:

Theorem 8.2.5 (Characterization of Capture Basins). *Let us assume that F is Marchaud, that the environment $K \subset X$ and the target $C \subset K$ are closed subsets satisfying $K \setminus C$ is a repeller ($\mathrm{Viab}_{\mathcal{S}}(K \setminus C) = \emptyset$) and K is backward invariant.*

*Then the viable capture basin $\mathrm{Capt}_{\mathcal{S}}(K, C)$ is the **unique** closed subset D satisfying $C \subset D \subset K$ and*

$$\begin{cases} (i)\ D \backslash C \text{ is locally viable under } \mathcal{S} \\ (ii)\ D \text{ is backward invariant relatively to } K \text{ under } \mathcal{S} \end{cases} \tag{8.6}$$

8.2.3 Tangent and Normal Cones

Let $x \in K \subset X$. The *tangent cone* to K at x is defined by

$$T_K(x) := \left\{ v \in X \mid \liminf_{h \mapsto 0+} \frac{d(x + hv; K)}{h} = 0 \right\}$$

Denoting by

$$P^{\star} := \{ p \in X^{\star} \text{ such that } \forall\, v \in P,\ \langle p, v \rangle \leq 0 \}$$

the polar of P,

- the normal cone $N_K(x) := T_K^\star(x)$
- $T_K^{\star\star}(x)$ is the closed convex hull of $T_K(x)$.

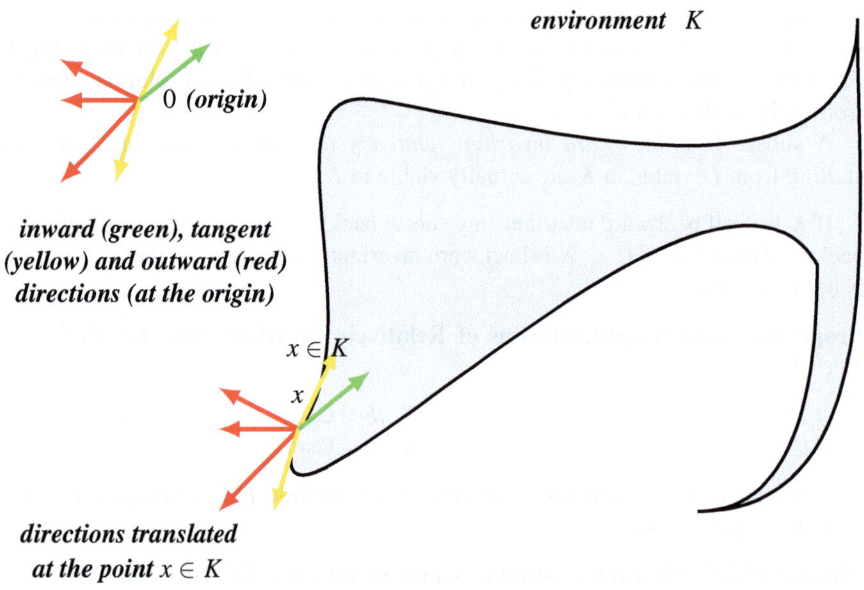

38 (Schematic Representation of Tangent Cones) *We represent the environment K, an element $x \in K$ and the origin. Six vectors v are depicted: one which points inward K, and thus tangent to K, two tangent vectors which are not inward and three outward vectors. Their translations at x belong to K for the inward vector, "almost" belong to K for the two tangent and not inward vectors and belong to the complement of K for the three outward vectors.*

8.2.4 The Regulation Map

These theorems, which are valid for any evolutionary systems, paved the way to go one step further when the evolutionary system is a differential inclusion.

We shall use the closed convex hull $T_K^{\star\star}(x)$ of the tangent cone.

Not only the Viability Theorem provides characterizations of viability kernels and viable-capture basins, but also the *regulation map* $R_\Delta(x) \subset U(x)$ which governs viable evolutions:

Definition 8.2.6 (Regulation Map). Let us consider three subsets $C \subset D \subset K$ (where the target C may be empty) and a control system (f, U).

The set-valued map R_Δ defined by

$$\forall\, x \in D \setminus C,\quad R_\Delta(x) \;:=\; \{u \in U(x) \text{ such that } f(x,u) \in T_D(x)\} \tag{8.7}$$

is called the regulation map of (f, U) on $D \setminus C$.

The Viability Theorem states that

Theorem 8.2.7 (Viability Theorem). *Let us assume that (f, U) is Marchaud, that the environment $K \subset X$ and the target $C \subset K$ are closed subsets such that $K \setminus C$ is a repeller ($\mathrm{Viab}_F(K \setminus C) = \emptyset$). Then the viable-capture basin $\mathrm{Capt}_{\mathcal{S}}(K, C)$ is the* ***largest*** *closed subset D satisfying $C \subset D \subset K$ and*

$$\forall x \in D \setminus C,\quad R_\Delta(x) \;\neq\; \emptyset$$

Furthermore, for every $x \in D$, there exists at least one evolution $x(\cdot) \in \mathcal{S}(x)$ viable in D until it reaches the target C and all evolutions $x(\cdot) \in \mathcal{S}(x)$ viable in D until they reach the target C are governed by the differential inclusion

$$x'(t) \;\in\; R_\Delta(x(t))$$

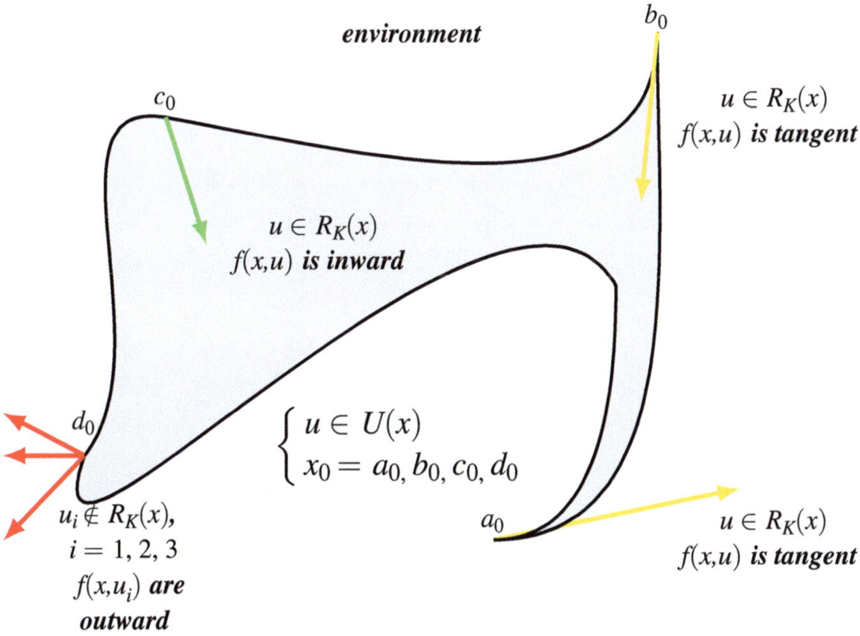

39 (Schematic Description of the Regulation Map) *In this scheme, we describe four situations at elements* a_0, b_0, c_0 *and* $d_0 \in K$. *At* a_0 *and* b_0, *the velocities* $f(x,u)$ *are tangent to* K, *so that we can expect an evolution to be viable. At* c_0, *this hope is even more justified because the velocity* $f(x,u)$ *points in the interior of* K. *Finally, at* d_0, *all velocities* $f(x,u)$ *point outward* K, *and it is intuitive that all evolutions leave* K *instantaneously. The viability theorem states that these intuitions and hopes are correct for any closed subset* K *and for Marchaud maps.*

Chapter 9
What Is to Be Done?

Shto delat? This title of the Nikolai Chernyshevsky' 1863 novel became famous in 1901 when Lenin took over this tile.

The answer is that everything remains to be done, since life sciences, and, among them, economics, escape, for a large part, our understanding.

Restricting our interrogations to mathematical metaphors, it is necessary to flush out the devil that lurks in the mathematical details and to understand why and how he is adept at digging other hiding places in the world to conceal our cultural understanding of other details. Throughout this interminable game of Sisyphus, scholars are trying to dry the soil where this devil buries our conceptual treasures, but so slowly that it retains all its chances of remaining eternal.

However, some open problems could be investigated with some of the available tools of Viability Theory, hoping that they will soon trigger motivations to forge new ones, probably radically different.

9.1 Insuring Each Citizen

Each agent should be guaranteed to receive a minimum income based on several parameters: time t, age a, health h, income i, population p, employment e, etc. For example, regarding the dependency of age, the income may be positive for juniors (cost of their education, for example) and for seniors (pensions) and negative for "active" (for example, in the form of taxes or contributions, etc.). These suggestions are proposed not only for ethical reasons and fairness purposes, but also for the economic system to run. This more general endowment problem involves more variables than the duration, commodities, and prices we presented above. However, the same mathematical theorems can be used. The real difficulties to overcome are political, on one hand, and technological, on the other hand, since important computational means would be necessary.

J.-P. Aubin, *Time and Money*, Lecture Notes in Economics and Mathematical Systems 670, DOI 10.1007/978-3-319-00005-3_9,

These constraints could be described or summarized in the form of a threshold function $\mathbf{k} : (t, a, h, i, p, e, \ldots) \mapsto \mathbf{k}(t, a, h, i, p, e, \ldots)$ depending on the chosen relevant variables.

Hence, the question is to endow each citizen with an endowment $y(\cdot)$ satisfying

$$\forall\, t \in [T - \Omega, T], \;\; y(t) \;\geq\; \mathbf{k}(t, a(t), h(t), i(t), p(t), e(t), \ldots)$$

Actually, the question relies on the determination of the dynamics governing the future evolutions of these variables. They are not really unknown. However, we need *to know only the worst case evolutions*, i.e., the *smallest forecasted velocities of these variables extrapolated from the knowledge of historical time series* for insuring this threshold value. The "model risk" lies at this stage, which is an assumption, so that the conclusions are mathematically (and thus, rigorously) deduced from these assumptions.

This is the "tychastic approach" used for insuring portfolios in finance (see [17, *Tychastic Measure of Viability Risk. A Viabilist Portfolio Performance and Insurance Approach*] offers an introduction to these methods for the uncertain variables time and asset prices only).

The evolutions of these parameters are indeed uncertain: dates of birth and death, health or hardship (measured by the cost of care, for instance), income (salary, heritage, etc.), needs (children), total population, employment (and unemployment), and many other variables.

Two methods (at least) can be considered: one eradicating the risk according to forecasts and deduce the insurance, the other, assuming that the threshold given and computing the worst case velocities:

1. *The predictive approach*:

    a. *Assumptions*: At each date, the worst possible velocities (or rates) of the variables involved by methods of extrapolation or prediction are available during the remaining temporal window;
    b. *Conclusions*: At each date, the (minimal) guaranteed endowment allowing the threshold to be hedged during the remaining temporal window (see Chap. 3, p. 47).

2. *The impulsive approach*:

    a. *Assumptions*: the threshold and a guaranteed endowment and/or velocity is accepted whenever the endowment reaches the threshold (see Sect. 2.3, p. 37);
    b. *Conclusions*: At each date, the worst velocities or rates of the variables allowing this scenario are computed.

These two approaches are inverse: in the predictive approach, the future is anticipated and the insurance is computed, whereas in the impulsive approach, we do not forecast the future but we impose a rule to respond to disasters: if the actual income is equal to the threshold, the "insurer" provides the insured the guaranteed endowment. We then calculate the consistent uncertainty necessary for this impulsive mechanism to be operational.

Tychastic viability theory allows us to solve mathematically these problems. They differ from stochastic or probabilistic techniques, which provide *statistical evaluations*: mean, variance, values at risk, expected shortfalls (assumed to measure the risk), and other such methods. *The risk is not eliminated nor mitigated, but only estimated.*[1] These currently used techniques do not provide the computation of neither equity (or solvency capital requirement, etc.) nor the associated management rule which guarantee that the value of the portfolio is always above the liability.

9.2 Kingly Duties

The fundamental duty of a sovereign State is to guarantee its citizens the various components of their security: external security (military, diplomatic); homeland security (justice, police); social security (tax and redistribution); etc.

We addressed the social protection aspects to investigate mathematical methods for solving a tiny aspect of this problem. Mathematics are far to bring solutions, and it is necessary to clarify the policy choices in order to avoid misunderstandings and misinterpretations.

1. *State or "Market"?* The mathematical investigation provides methods for insuring the citizens throughout his life, whether it is guaranteed by the State or the "Market." It is customary to contrast the collective investments implemented by the State and those by private enterprizes (pension funds, life insurance, etc.) or the "Market." To the extent that duration plays a crucial role in long-term insurance, it seems preferable to entrust its management to the States, which almost never die, instead of private, subjective to successive failures on shorter terms of companies as prone to "creative destruction" according to *Joseph Schumpeter.*
2. *Contributory or funded pension funds?* The debate on the comparative advantages between the contributory (or "unfunded") and funded pension funds systems is made confused by the role of contributory pensions funds (age-structured pay-as-you-go formulas) reserved to the States and the funded ones to private companies. The State may, if he has the will, use also a collective funding mechanism for social protection, even by using market mechanisms. Private funding focuses on time, to ensure the pension of each person at the end of life, whereas the contributory is based on an intergenerational context.

[1] And curiously, recommended by the Basel committees as these techniques were the only ones known to be available (and not improvable in the future) by their members. Laws should not incorporate scientific results and methods which evolve faster and become obsolete.

3. *Taxes or investment?* The issue of financing social protection is upstream of the proposed study. Once the guarantee is calculated, the funding mechanism has to be investigated. It can be funded by the States, with their redistributive and pooling effects, or by individual investments.
4. *Individual or collective plans?* Since *Bismark* and *Beveridge*, the building of the welfare state pension plans took over systems that favored individual pension for the richest individuals and depriving the poorest of insurance except from their progeny. States have invented and designed progressively pension mechanisms to each social groups, so numerous and different that the bush they form favors a detrimental opacity and unfairness.
5. *Threshold effects or continuous management*: a source of problems are due to "threshold effects," cutting the variables (duration, age, income, etc.) in a small number of "slices" (intervals) for facilitating their management. The arbitrariness of these slices introduces perverse incentives that could be mitigated by a more continuous time management (with more numerous and smaller slices), that are allowed nowadays thanks to more powerful digital computing power.

9.3 Is It Time to Conclude?

Talking conclusion is to admit that a goal is proposed, while this study opens just problems that we do not know whether they are relevant, or whether they will be resolved. To paraphrase Paul Valery about life, *"if science had a goal, it is not science"*.

Rather than a goal, this study is motivated by a financial crisis driven by "short-termist" greed of a tiny minority of predators that thrive on poverty of the vast majority of human beings, not to mention destroying natural and not-renewable resources. If we judge harmful unjustified financial fluidity, which is an hypothesis that can be accepted or not, so the urges of the speculators should be estimated and computed for avoiding excessive profits, instead of only slowing down without justification the transactions of economic agents, so that the financial investment brings to fruition its economic fruits.

The "Outraged" or "Occupy Wall Street" movements and economists sharing their concerns are blamed for their contemptible ignorance or disbelief in the faith to "Market" by speculators and those they control by corruption and swollen by the voices of the media. Instead of burning these dissidents in the public square, they are starved in dribs and drabs. In the minds of speculators and of their supporters, they get what they deserve, their misery, and so worse if they are not responsible. In their opinion, the workers were and still are always guilty for not working and producing enough, and their complaints do not reach the ears of autistic rulers more afraid of the speculators than of the people they govern. They prefer to borrow state money from private banks rather than from their fellow citizens[2] or their central banks. They are more afraid of rates awarded by "self-proclaimed" credit rating agencies (that they support!) than by all those who, in the least bad cases, have elected them to govern their destiny in their name.

What is to be done? The "inertia principle" (see Box 21, p. 58) teaches that only when the situation reaches the viability frontier, and not before, the first available viable solution is chosen after much trials and errors. States will likely default. Having failed to restructure a reasonable part of the excessive interest of the debt, and to boost the economy by the adequate monetary endowment needed for maintaining viable the economy, they will inevitably be ruined by austerity policies. Privileging speculative investment over productive investment, they deprive of work those they turn into "devalued" persons, both economically and psychologically, resulting in general recession, and thus, defaulting anyway, with direr consequences.

The duty of politicians is to anticipate viable solutions to correct intelligently past mistakes and avoid this announced disaster. The true courage is to ensure the best interests of the majority, and the ones of a minority of speculators, however powerful they are. This is matter of will and confidence in his own people.

Governing is understanding, before forecasting, planning, choosing,[3] and acting.

[2] As the Belgium government did in November 2011 with an unexpected success.

[3] *"Gourverner, c'est prévoir"*, according to *Émile de Girardin*, a French journalist from the XIXth century, *"Gouverner, c'est choisir"* is the title of one of the books by *Pierre Mendès France*, who served as the Prime Minister of France for 7 months and 17 days (1954–1855), leaving an imprint inversely proportional to the duration of his power. Pierre Mendès France was an honest and profound intellectual, yet an economist, a social planner, and, in our view, one of the best French afterwar statesman. See *Jean Lacouture* [98, *Pierre Mendès-France*] for a biography.

References

1. Amable B (2005) Les cinq capitalismes : Diversité des systèmes économiques et sociaux dans la mondialisation. Seuil
2. Andersen H, Grush R (2009) A brief history of time-consciousness: historical precursors to James and Husserl. J Hist Philos 47:277–307
3. Anita S (2000) Analysis and control of age-dependent population dynamics. Kluwer Academic
4. Arrow K, Hahn F (1971) General competitive analysis. Holden Day
5. Aubin J-P (1979, 1982) Mathematical methods of game and economic theory. North-Holland (Studies in Mathematics and its applications, vol 7, pp 1–619), reprint by Dover (2007)
6. Aubin J-P (1981) A dynamical, pure exchange economy with feedback pricing. J Econ Behav Organ 2:95–127
7. Aubin J-P (1991) Viability theory. Birkhäuser
8. Aubin J-P (1996) Neural networks and qualitative physics: a viability approach. Cambridge University Press, Cambridge
9. Aubin J-P (1997) Dynamic economic theory: a viability approach. Springer
10. Aubin J-P (1998, 1993) Optima and equilibria. Springer. French Version: (1984) L'analyse non linéaire et ses motivations économiques. Masson
11. Aubin J-P (2000) Mutational and morphological analysis: tools for shape regulation and morphogenesis. Birkhäuser
12. Aubin J-P (2010) La mort du devin, l'émergence du démiurge. Essai sur la viabilité, la contingence et l'inertie des systèmes. Beauchesne
13. Aubin J-P (2011) Viability solutions to structured hamilton–jacobi equations under constraints. SIAM J Contr Optim 49:1881–1915. http://dx.doi.org/10.1137/10079567X
14. Aubin J-P (2011) Regulation of births for viability of populations governed by age-structured problems. J Evol Equat 99–117, DOI:http://dx.doi.org/10.1007/s00028-011-0125-z
15. Aubin J-P, Bayen A, Saint-Pierre P (2011) Viability theory. New directions. Springer
16. Aubin J-P, Bonneuil N, Maurin F (2000) Non-linear structured population dynamics with co-variables. Math Popul Stud 91:1–31
17. Aubin J-P, Chen LX, Dordan O (in preparation) Tychastic measure of viability risk. A viabilist portfolio performance and insurance
18. Aubin J-P, Chen LX, Dordan O (2012) Asset liability insurance management (ALIM) for risk eradication. In Bernhard P, Engwerda J, Roorda B, Schumacher H, Kolokoltsov V, Saint-Pierre P, Aubin J-P (eds) The interval market model in mathematical finance. Game-theoretic methods. Birkhäuser
19. Aubin J-P, Chen LX, Dordan O, Saint-Pierre P (2011) Viabilist and tychastic approaches to guaranteed ALM problem. Risk Decis Anal 3:89–113, DOI:10.3233/RDA-2011-0033

J.-P. Aubin, *Time and Money*, Lecture Notes in Economics and Mathematical Systems 670, DOI 10.1007/978-3-319-00005-3,

20. Aubin J-P, Chen LX, Dordan O, Faleh A, Lezan G, Planchet F (2012) Stochastic and tychastic approaches to guaranteed ALM problem. Bull Français d'Actuariat 12:59–95
21. Aubin J-P, Cellina A (1984) Differential inclusions. Set-valued maps and viability theory. Springer
22. Aubin J-P, Dordan O (1996) Fuzzy systems, viability theory and toll sets. In Nguyen H(ed) Handbook of fuzzy systems, modeling and control. Kluwer, pp 461–488
23. Aubin J-P, Dordan O (in preparation) Evaluation and quotations of sets
24. Aubin J-P, Frankowska H (1990) Set-valued analysis. Birkhäuser
25. Aubin J-P, Haddad G (2001) Path-dependent impulse and hybrid control systems. In Di Benedetto, Sangiovanni-Vincentelli (eds) Hybrid systems: computation and control, pp 119–132. Proceedings of the HSCC 2001 Conference, LNCS 2034. Springer
26. Aubin J-P, Haddad G (2002) History (Path) dependent optimal control and portfolio valuation and management. J Positivity 6:331–358
27. Barro RJ, Grossman HI (1971) A general disequilibrium model of income and employment. Am Econ Rev 61:82–93
28. Barron EN, Jensen R (1990) Semicontinuous viscosity solutions for Hamilton–Jacobi equations with convex Hamiltonians. Comm Part Differ Equat 15:1713–1742
29. Blanchflower DG, Oswald A (2004) Money, sex and happiness: an empirical study. Scand J Econ 106:393–415
30. Bénassy JP (1993) Nonclearing markets: microeconomic concepts and macroeconomic applications. J Econ Lit 31:732–761
31. Bénassy JP (2002) The macroeconomics of imperfect competition and non-clearing markets. MIT Press
32. Bergson H (1903) Introduction à la métaphysique [English translation: The creative mind: an introduction to metaphysics. Replica Books (2001)]
33. Bergson H (1907) L'évolution créatrice [English translation: Creative evolution. Dover (1998)]
34. Bewley TF (2007) General equilibrium, overlapping generations models, and optimal growth theory. Harvard University Press
35. Bohm von Bawerk E (1884) History and critique of interest theories, in Capital and interest (3 volume set). Libertarian Press (1959)
36. Bos A (1891) Glossaire de la langue d'oïl, (XI-XIV siècles). Maisonneuve (Book digitized by Google from the library of Harvard University, http://archive.org/details/glossairedelala00bosgoog)
37. Brahic E, Terreaux J-P (2009) Évaluation économique de la biodiversité. Quae
38. Brosnan SF, Jones, Gardner M, Lambeth SP, Schapiro SJ (2012) Evolution and the expression of biases: situational value changes the endowment effect in chimpanzees. Evol Hum Behav 33:378–386
39. Broad CD (1951) A logistic analysis of the two-fold time theory of the specious present. Br J Philos Sci 2:137–141 (Appendix to [57, Dobbs])
40. Butler FM (1935) Lincoln money martyred. Lincoln Publishing (http://www.archive.org/stream/lincolnmoneymart00butl/lincolnmoneymart00butl_djvu.txt)
41. Cartelier J (1996) La monnaie, Flammarion
42. Cartelier J (2002) Monnaie ou don: réflexion sur le mythe économique de la monnaie. J des Anthropologues 90–91:353–374
43. Cartelier J (2006) Comptabilité et pensée économique: Introduction à une réflexion théorique. Revue Economique 57(5):1009–1032
44. Cartelier J (2007) The hypostasis of money: an economic point of view. Cambridge J Econ 31:217–233
45. Cartelier J (2011) Money and sovereignty: a comparison between Hobbes and modern money theory. In: Heiner Ganssmann, Routledge (eds) New approaches to monetary theory. Routledge International Studies in Money and Banking
46. Cartelier J (2011) Money is the scribe of a market economy. In: Amato M, Doria L, Fantacci L (eds) Money and calculation. Economic and sociological perspectives. Macmillan, New York

47. Champsaur P, Drèze G, Henry C (1974) Dynamic processes in economic theory, vol 7417, de Center for Operations Research and Econometrics Leuven
48. Champsaur P, Drèze G, Henry C (1977) Stability theorems with economic applications. Econometrica 45:273–294
49. Cornet B (1977) An abstract theorem for planning procedures. Lect Notes Econ Math Syst 144:53–59
50. Cornet B (1977) Accessibilité des optima de Pareto par des processus monotones. Comptes Rendus de l'Académie des Sciences 285:641–645
51. Cornet B (1978) On planning procedures defined by multivalued differential equations. Systèmes Dynamiques et Modèles Economiques Colloque International du C.N.R.S 259:111–122
52. Cornet B (1979) Monotone planning procedures and accessibility of Pareto optima. In: Aoki et Marzollo (eds) New trends in dynamic system theory and economics. Academic Press, pp 337–349
53. Cornet B (1983) Neutrality of planning procedures. J Math Econ 11:141–160
54. Cornet B (1990) Neutrality of planning procedures. J Math Econ 11:141–160
55. Debreu G (1959) Theory of value. Wiley
56. De Vroey M (2004) Involuntary unemployment: the elusive quest for a theory. Routledge
57. Dobbs HAC (1951) The relation between the time of psychology and the time of physics part I. Br J Philos Sci 2(6):122–141
58. Dobbs HAC (1951) The relation between the time of psychology and the time of physics part II. Br J Philos Sci 2(7):177–192
59. Dobbs HAC (1969) The present in physics. Br J Philos Sci 19:317–324
60. Dordan O (1995) Analyse qualitative. Masson
61. Drèze G, de la Vallée Poussin D (1971) A tâtonnement process for public goods. Rev Econ Stud 38:133–150
62. Drucker P (1973) The practice of management. Colins
63. Einzig P (1948) Primitive money. Eyre & Sopttiwoode
64. Evans LC (1998) Partial differential equations. American Mathematical Society
65. Flürscheim M (1909) The economic and social problems. Jefferson Publishing
66. Frankowska H (1991) Lower semicontinuous solutions to Hamilton–Jacobi–Bellman equations. In: Proceedings of the 30th IEEE conference on decision and control, Brighton, UK
67. Frankowska H (1993) Lower semicontinuous solutions of Hamilton–Jacobi–Bellman equations. SIAM J Contr Optim 31:257–272
68. Frankowska H (2005) Optimal synthesis via superdifferentials of value function. Contr Cybern 34:787–803
69. Frankowska H (2010) Control under state constraints. In: Proceedeings of the International Congress of Mathematicians (ICM 2010), Hyderabad, India
70. Fuchs G (2007) Dépasser le capitalisme. L'Harmattan
71. Galbraith JK (1958) The affluent society. Houghton Mifflin
72. Gale D (1983) Money in disequilibrium. Cambridge University Press, Cambridge
73. Giraud G, Renouard C (2012) Vingt propositions pour réformer le capitalisme. Flammarion
74. Goebel R, Sanfelice RG, Teel AR (2012) Hybrid dynamical systems. modeling, stability and robustness. Princeton University Press, Princeton
75. Gordon D (2011) The turgot collection. Ludwig von Mises Institut
76. Gould SJ (1987) Time's arrow, time's cycle. Harvard University Press
77. Grandsaignes d'Hauterive R (1947) Dictionnaire d'ancien français. Moyen-Age et Renaissance. Larousse
78. Griffon M, Griffon L (2011) Pour un monde viable: changement global et viabilité planétaire. Odile Jacob
79. Grünbaum A (1969) Are physical events themselves transiently past, present and future? A reply to H. A. C. Dobbs. Br J Philos Sci 20:145–153
80. Haddad G (1981) Monotone trajectories of differential inclusions with memory. Isr J Math 39:83–100

81. Haddad G (1981) Monotone viable trajectories for functional differential inclusions. J Differ Equat 42:1–24
82. Haddad G (1981) Topological properties of the set of solutions for functional differential inclusions. Nonlinear Anal Theory Meth Appl 5:1349–1366
83. Hahn F, Negishi T (1962) A theorem on non-tâtonnement stability. Econometrica 30:463–469
84. Hanemann MW (1991) Willingness to pay and willingness to accept: how much do they differ? Am Econ Rev 81:635–647
85. Harbaugh WT, Krause K, Vesterlund L (2001) Are adults better behaved than children? Age, experience, and the endowment effect. Econ Lett 70(2):175–181
86. Hayek F (1988) The fatal conceit. Routledge, Briggs
87. Heichelheim FM (1958) An ancient economic history. Sijthofs Ultgeversmaatschappij 1:21
88. Hildenbrand W, Kirman AP (1988) Equilibrium analysis. North-Holland
89. Homer S, Sylla RE (1996) A history of interest rates. Rutgers University Press
90. Iannelli M (1995) Mathematical theory of travel-time-structured traffic management. Giardini, Pisa
91. James W (1890) Principles of psychology, 2 vols. Henry Holt
92. Kanngiesser P, Santos LR, Hood BM, Call J (2011) The limits of endowment effects in great apes (Pan paniscus, Pan troglodytes, Gorilla gorilla, Pongo pygmaeus). J Comp Psychol 125:436–445
93. Kelly ER (1882) The alternative: a study in psychology. Macmillan
94. Keyfitz N, Caswell H (2005) Applied mathematical demography, 3rd edn. Springer Series: Statistics for Biology and Health
95. Keyfitz B, Keyfitz N (1997) The McKendrick partial differential equation and its uses in epidmiology and population study. Math Comput Model 26:1–9
96. Krehl POK (2008) History of shock waves, explosions and impact. Springer
97. Krugman P (2012) End this depression now!. Norton & Company
98. Lacouture J (1981) Pierre mendès-France. Le Seuil
99. Larrouturou P (2011) Pour éviter le krach ultime. Nova editions
100. Lordon F (2010) Capitalisme, désir et servitude. La Fabrique éditions
101. Lordon F (2011) L'intérêt souverain - Essai d'anthropologie économique. La Découverte, 2011
102. Larrouturou P (2007) Le livre noir du libéralisme. Editions du Rocher (préface de Michel Rocard)
103. Longo G, Montévil M (2013) Biological time, symmetries and singularities. Springer
104. Lorenz T (2010) Mutational analysis. A joint framework for cauchy problems in and beyond vector spaces, series: lecture notes in mathematics, vol. 1996. Springer
105. Malinvaud E (1971) A planning approach to the public good problem. Swed J Econ 73:96–112
106. Maupertuis P-L Moreau de (1748) Réflexions philosophiques sur l'origine des langues et la signification des mots
107. Maupertuis P-L Moreau de (1745) Vénus physisque
108. Maupertuis P-L Moreau de (1744) Dissertation physique à propos du nègre blanc
109. Maupertuis P-L Moreau de (1751) Système de la nature
110. Maupertuis P-L Moreau de (1756) à Essay de philosophie morale
111. Neary JP, Stiglitz JE (1983) Toward a reconstruction of Keynesian economics: expectations and constrained equilibria. Q J Econ 98:199–228
112. Ostroy JM, Starr RM (1974) Money and the decentralization of exchange. Econometrica 42:1093–1113
113. Roche M (2010) La Banque. comment goldman sachs dirige le monde. Albin Michel
114. Rockafellar RT, Wets R (1997) Variational analysis. Springer
115. Sapir J (2011) La démondialisation. Seuil
116. Smale S (1876) Dynamics in general equilibrium theory. Am Econ Rev 66:88–94
117. Soddy F (1934) The rôle of money. Routledge. http://archive.org/details/roleofmoney032861mbp

118. Solow RM, Stiglitz JE (1968) Output, employment and wages in the short run. Q J Econ 82:537–560
119. Stachurski J, Venditti A, Yano M (2012) Nonlinear dynamics in equilibrium models: chaos, cycles and indeterminacy. Springer
120. Stiglitz JE (2012) The price of inequality: how today's divided society endangers our future. W.W. Norton & Company
121. Uzawa H (1962) On the stability of Edgeworth's Barter process. Int Econ Rev 3:218–232
122. Von Foerster H (1959) Some remarks on changing populations. In: The kinetics of cell proliferation, pp 382–407
123. Von Neumann J, Morgenstern 0 (1944) Theory of games and economic behaviour. Princeton University Press, Princeton
124. Webb G (1985) Theory of nonlinear age-dependent population dynamics. Marcell Dekker
125. Weber J (2010) L'évaluation contingente: les valeurs ont-elles un prix? In: http://filex.cirad.fr/get?k=z1nIvEVUoL2YBRdZBi7
126. Weigel E (Erhardi VVeigelii) (1669) Idea matheseos universæ cum speciminibus inventionum mathematicarum
127. Wiggins S (1990) Introduction to applied nonlinear systems and chaos. Springer

Index

Advised commodity, 108
Advised global commodity, 113
Anticipation, 80
Aperture, 32, 68, 69
Aperture functional, 71
Aperture of the temporal window, 70
Auspicious, 1, 68

Beginning of the temporal window, 71
Big Mac Index, 5

Calendar duration, 69, 72
Canonicals dynamics, 111
Capture basin, 120
Capture basin of C viable in K, 120
Chaperon, 69
Characteristic system (reverse), 50
Chow Mein Index, 5
Chronological time, 68
Clio derivative, 81
Closed loop controls, 120
Commodities effect, viii
Commodity space, 105
Concatenation, 78
Configuration space, 105
Constant duration, 72
Constant fluidity duration, 72
Constraint cost function, 110
Consumer price indexe, 5, 108
Contingent map, 119
Contingent reservoir, 119
Critical zone, 62
Cyclic evolution(s), 63
Cyclic evolutionary systems, 63

D:TMEndowmentFunction2, 88, 112
Departure set, 87
Deterministic evolutionary system, 79, 80
Differential inclusion, 37
Direct approach, 21, 33
Dirichlet boundary condition, 41
Disviability, 25
Domain, 69
Durance, 70
Duration function, 70, 73, 106
Dynamic optimality, 90, 98, 115

Economic tube, 87
Economical clocks, 63
Ending time, 71
Endowment, 2, 4, 32
Endowment effect, 2, 96
Endowment function, 49, 51, 89
Endowment fund, 2
Ergosophy, 16
Evolution, 68
Evolutionary system, 119
Evolutions with bounded fluidity, 73
Exchange value, 12
Exponential inertia function, 61
Extrapolator, 80

Feedback, 33, 120
Fluctuation (price), 106
Fluidity of duration function , 71
Flux, 113
Force, 105
Fundamental value, 12
Fuzzy economy, 110

J.-P. Aubin, *Time and Money*, Lecture Notes in Economics and Mathematical Systems 670, DOI 10.1007/978-3-319-00005-3,

Global commodity, 107

Haste, 75
Heavy evolution, 59, 60
Hysteresis, 66

Impetuosity, 75, 107, 113
impetuosity, 106
Impetus, 4, 32, 36, 88, 106–108
Impulse, 36
Impulse dynamical systems, 36
Impulsiveness, 75
Inertia function, 60
Inertia principle, 58
Inertia threshold, 58, 62, 63
Inflation, 5, 108
Instant, 69
Instantaneous cost function, 111
Instantaneous duration, 72
Inverse approach, 21
Irreversibility, 80
Isolated, 120

Kairos, 62

Logistic, 33
Lower semicontinuous, 111

Management rule, 33
Marchaud map, 122
Maximal impetus, 4

Newtonian inertia function, 60
Numéraire, 2

Ockham, 23
Open loop controls, 120
Optimal temporal window, 4
Optimality threshold, 91

Pantometria, 11
Patrimonial value, 4, 36, 37, 105
Perception of time, 74
Periodic evolution, 63
Potential function, 54, 87, 107, 112
Power, 105, 107
Precipitation, 75
precipitation, 106
Prediction, 80
Present time, 68
Price, 105, 107, 113
Price effect, viii
Price fluctuation, 36, 105
Price fluctuation impact, 4
Private affluence and public squalor, 8
Propitious, 68
Punctuated equilibrium, 58, 59

Rate of return, 49
Recalling operator, 81
Recording operator, 81
Redondance, 119
Redundancy, 80
Regulation law, 33, 117
Regulation map, 33, 58, 80, 124
Repeller, 120, 123
Return, 47
(Reverse) characteristic control system, 89, 99
(Reverse) characteristic system, 50
Reverse discount factor, 49
Riba, 17
Ribbit, 17

Sluggish growth, 65
Solvency capital requirement, 48
Speed limit, 63
Sundial, 77

Tâtonnement rule, 4
Temporal window, 86
Temporal window of a duration function, 71
Threshold commodity function, 41
Time acceleration, 73
Time reverser, 78
Time-duration map, 110
Time-duration pair, 70
Timing, 62
Titanic effect, 64, 65
Trajectory, 68
Transaction, 32, 105, 106
Transaction map, 33
Transaction potential, 88
Transaction rule, 4
Transactions value, viii, 4
Tyches, 80

Uconomia, 24
Urge, 106, 107, 113
Urge (to act), 75

Value, 32
Value pairing, 105
Viability cycle, 66
Viability detector, 16
Viability function, 89, 99
Viability kernel, 120
Viability kernel of K with target C, 120
Viability property, 25, 32
Viability threshold condition, 31, 35
Viability threshold endowment function, 37
Viability volution, 51, 89, 99
Viable, 110
Viable outside a target, 120

Warning signals, 5, 62
Warning time, 62
Work, 105

Index of Names

Aldrich (Nelson Wilmarth) [], 7
Alembert (Jean Le Rond d') [1717–1783], 27
Amable (Bruno) [], 3
Arrow (Kenneth) [1921-], 27
Augustine (bishop of Hippo) [[354–430]], 76
Augustine (of Hippo) [354–430 BC], 68
Aumann (Robert) [1930-], 27

Böhm von Bawerk (Eugen) [1851–1914], 10
Bellman (Richard Ernest) [1920–1984], 90
Bergson (Henri) [1859–1941], 77
Beveridge (William Henry) [1879–1963], 130
Bismark (Otto Von) [1815–1898], 130
Blanchflower (David Graham) [1952-], 12
Boileau-Despréaux (Nicolas) [], 67
Borda (Jean-Charles, chevalier de) [1733–1799], 27
Bos (Alphonse) [1835–1913], 70
Brouwer (Luitzen) [1881–1966], 27

Canard (Nicolas-François) [1750–1838], 27
Carathéodory (Constantin) [1873–1950], 90
Chen (Luxi) [1987-], xi
Chernyshevsky (Nikolay Gavrilovich) [1828–1889], 127
Clément-Pitiot (Hélène) [1962-], xi, 16
Condorcet (Marie Caritat, marquis de) [1743–1794], 27
Cournot (Augustin) [1811–1877], 27

Désilles (Anya) [1972-], xi
Debreu (Gérard) [1921–2004], 27
Dobbs (H. A. C.) [], 68
Dordan (Olivier) [1962-], 12
Drucker (Peter) [1909–2005], 19
Duhamel du Monceau (Henri-Louis) [1700–1782], 9

Eldredge (Nils) [1943-], 58
Euler (Leonhard) [1707–1783], 75
Evans (Lawrence Craig) [1949-], 41
Ewing (Sir James Alfred) [1855–1935], 66

Fan (Ky) [1914–2010], 27
Flürscheim (Michael) [1844–1912], 17
Fourastié (Jean) [1907–1990], 5
Franklin (Benjamin) [1706–1790], 9
Fuchs (Gérard) [1940-], 3

Galbraith (John Kenneth) [1908–2006], 8
Galperin (Efim A.) [1934-], 32, 70
Girardin (Émile de) [1806–1881], 131
Giraud (Gaël) [1970-], 2, 3
Goebel (Rafal) [], 36
Gould (Stephen J.) [1941–2002], 58, 77
Grandsaignes d'Hauterive (Robert) [1881–1962], 70
Greenspan (Alan) [1926-], 18

Haddad (Georges) [1951-], xi
Hamilton (Sir William Rowan) [1805–1865], 75
Hayek (Friedrich) [1899–1992], 11, 27
Heinroth (Oskar) [1871–1945], 76
Hicks (John R.) [1904–1989], 27
Hume (David) [1711–1776], 9, 12, 85

J.-P. Aubin, *Time and Money*, Lecture Notes in Economics and Mathematical Systems 670, DOI 10.1007/978-3-319-00005-3,

Isaacs (Rufus Philip) [1914–1981], 90

Jacobi (Carl Gustav Jacobi) [1804–1851], 75
James (William) [1842–1910], 68
Juglar (Joseph Clément) [1819–1905], 63

Kakutani (Shizuo) [1991–2004], 27
Kelly (E. Robert), alias E.R. Clay [], 68
Kennedy (John Fitzgerald "Jack") [1917–1963], 7
Keynes (John Maynard) [1883–1946], 1, 27
Kondratiev (Nikolai) [1892–1938], 63
Krehl (Peter O.K.) [], 41
Krugman (Paul Robin) [1953-], 3, 23

Lévi-Strauss (Claude) [1908–2009], 5
Lacouture (Jean) [1921-], 131
Lagrange (Joseph Louis) [1736–1813], 75
Larrouturou (Pierre) [1964-], 3
Laskey (Ronald *Alfred*) [1945-], 10
Lenin (Vladimir Ilitch Oulianov) [1870–1924], 127
Lericolais (Nadia) [1966-], xi
Longo (Giuseppe) [], 70
Lordon (Frédéric) [1962-], 3
Lorenz (Thomas) [1974-], 20
Lysippos [c. 400 BC], 63

Mackay (Charles) [1814–1889], 29
Mairan (Jean-Jacques de) [1678–1771], 76
Malthus (Thomas) [1766–1834], 33
Marshall (Alfred) [1842–1924], 27
Martin (Sophie) [1977-], xi
Marx (Karl Heinrich) [1818–1883], 13
Maupertuis (Pierre-Louis Moreau de) [1708–1759], 74, 105
Mendès France (Pierre) [1907–1982], 131
Mirabeau (Victor Riqueti, marquis de Mirabeau) [1715–1789], 27
Mitterrand (François) [1916–1996], 75
Montévil (Maël) [], 70
Morgenstern (Oscar) [1902–1977], 27

Nash (John Forbes) [1928-], 27
Neumann (J. von) [1903–1957], 27
Nixon (Richard) [1913–1994], 5

Oswald (Andrew) [1953-], 12

Péguy (Charles) [1873–1914], 27
Pareto (Vilfredo) [1848–1923], 27
Plato [428–347 av. J.-C.], 9
Poincaré (Henri) [1854–1912], 11
Poisson (Siméon Denis) [1781–1840], 32
Ponzi (Charles) [1882–1949], 6

Quesnay (François) [1694–1774], 27

Renouard (Cécile) [1968-], 2, 3
Rocard (Michel) [1930-], 3
Roche (Marc) [1951-], 3
Rothschild (Nathan Meyer) [1777–1836], 17

Saint-Pierre (Patrick) [1946-], xi
Sapir (Jacques) [1954], 3
Scarf (Herbert) [1830-], 27
Schumpeter (Joseph Aloïs) [1883–1950], 129
Seemann (Karl-Henning) [1934-], vii
Shapley (Lloyd) [1923-], 27
Smith (Adam) [1723–1790], ix, 8, 9, 13
Soddy (Frederick) [1877–1956], 16
Spalding (Douglas Alexander) [1841–1877], 76
Stiglitz (Joseph Eugene) [1943], viii

Taylor (Edmund Dick) [1804–1891], 7
Terreaux (Jean-Philippe) [1961-], 68
Thaler (Richard) [1945-], 96
Tobin (James) [1918–2002], 19
Turgot (Anne Robert-Jacques) [1727–1781], 2, 12, 27, 85

Valéry (Paul) [1871–1945], 80
Verhulst (Pierre François) [1804–1849], 33
Vinter (Richard) [1948-], xi

Walras (Auguste A.) [1801–1866], 27
Walras (Léon) [1834–1910], 1, 8, 11, 25, 27
Washington (George) [1732–1799], 7
Welch (John Francis "Jack") [1935-], 18
Wiggins (Stephen) [1959-], 32
Wilson (Thomas Woodrow) [1856–1924], 7

MIX
Papier aus verantwortungsvollen Quellen
Paper from responsible sources
FSC® C105338

If you have any concerns about our products,
you can contact us on
ProductSafety@springernature.com

In case Publisher is established outside the EU,
the EU authorized representative is:
Springer Nature Customer Service Center GmbH
Europaplatz 3, 69115 Heidelberg, Germany

Printed by Libri Plureos GmbH
in Hamburg, Germany